THE TREMBLING
MOUNTAIN

A Personal Account of Kuru,
Cannibals, and Mad Cow Disease

THE TREMBLING
MOUNTAIN

A Personal Account of Kuru, Cannibals, and Mad Cow Disease

ROBERT KLITZMAN, M.D.

PERSEUS PUBLISHING

Cambridge, Massachusetts

Library of Congress Cataloging-in-Publication Data

Klitzman, Robert.
    The trembling mountain   a personal account of kuru, cannibals,
and mad cow disease / Robert Klitzman.
        p.   cm.
    Includes bibliographical references and index.
    ISBN 0-306-45792-X
    1. Prion diseases.   2. Kuru--Papua New Guinea.   3. Bovine
spongiform encephalopathy.   4. Medical anthropology--Papua New
Guinea.   I. Title.
    [DNLM: 1. Encephalopathy, Bovine Spongiform. 2. Kuru.
3. Cannibalism.   4. Anthropology.   WL 300 K656t 1998]
RA644.P93K57 1998
616.8--dc21
DNLM/DLC
for Library of Congress   98-14386
                                                                  CIP

All photos are courtesy of the author, except that on page 275,
which is courtesy of Jeanie Mckenzie.

ISBN 0-7382-0614-8

Published by Perseus Publishing
A Member of the Perseus Books Group

http://www.perseuspublishing.com/

10 9 8 7 6 5 4

Printed in the United States of America

From so simple a beginning endless forms most beautiful and most wonderful have been, are being, evolved.

—Charles Darwin
The Origin of Species

A man cannot lead a decent life in the cities, and I needed to live.

—Antoine de St. Exupery
Wind, Sand and Stars

Contents

III. BEYOND THE MOUNTAIN

IV. RETURNS

V. POSTSCRIPT

Preface

In 1996, when Mad Cow disease spread to humans, I sat at dinner parties and other gatherings as the topic came up and realized that people knew very little about this or related diseases. I had experience and firsthand knowledge that others were interested in hearing about pertaining to this new and strange epidemic. In particular, I had spent months in Papua New Guinea studying kuru, a disease caused by essentially the same infectious agent, which had wiped out much of the Stone Age Fore group there.

Kuru is now disappearing from the earth, but as the culprit, an infectious protein, resurfaces through Mad Cow disease, or bovine spongiform encephalopathy, the disease has become of increasing significance. In addition, the Stone Age fades ever further from the world. Whatever observations and firsthand accounts of it and the people who grew up in it now exist are all that ever will. Particularly as the world becomes increasingly modern and homogeneous, it is important to document as much as possible these all but lost roots of human culture.

These experiences also shed light on what it's like to do science, particularly field work in epidemiology and medical anthropology—the difficulties, ironies, and triumphs, and the ways in which scientists and anthropologists are made.

To protect confidentiality, I have changed certain details in this account.

PART I CROSSING TIME

Beefeaters

We had been hiking all day in the Yorkshire moors through mists, and over craggy grass that tightly gripped the light grey rocky ground. I kept lagging behind as we climbed the hills, along dried up river beds and waterfalls. At last we descended into a tiny valley reached by a road. By a stream stood a small pub, built entirely of the same grey stone on which we had been trekking all day. The year it had been built—1746—was painted in serifed gold letters over the black wooden lintel. Inside, a fire blazed in a hearth. My friends—all English—and I sat down and looked at menus. "What would you like?" a waitress asked.

My companions all ordered roast beef and Yorkshire pudding.

"Aren't you afraid of Mad Cow disease?" I asked them, astonished. A few years before, in 1985, cases of this disease, or bovine spongiform encephalopathy (BSE), had been found in British cattle. By 1988, hundreds of cows had become ill, and the British government banned feeding bone meal from the ground-up remains of sheep and cattle to other such animals as a protein supplement. (Sheep have long been known to have a related disease called scrapie.) The government later banned the marketing of sick cows to humans. In 1989, as the number of cases climbed into the thousands, the government barred selling animal parts most likely to be infected, notably brain and spinal cord, from infected herds. Still, thousands of cows died. In May 1990, a cat died of the disease, and British beef consumption fell. Yet it slowly rose back up. Now, in September 1990, I was amazed to see my friends all ordering beef.

"The beef is now safe," Mark, the other physician among us, answered, surprised at my question.

"But the infectious agent can take decades before it affects someone," I said. "If even one infected cow gets through, people could become infected and die."

"But the government has experts," Susan, a tall woman, argued. "They said the beef's okay." She lifted her fork and bit off a piece of pink meat. This willingness to accept government assurances shocked me, given the potential risks.

"I think it's better to be careful," I argued.

"Don't be silly," Susan said. "They've killed all the sick cows. Besides, we grew up eating beef. It's a British tradition."

"But the disease can take ten, maybe twenty years to show up or affect you." They all looked at me like I was crazy. I had an eerie feeling of something not right—of danger and blindness.

"It's never been transmitted to man," Mark replied.

"But it's been transmitted to other species."

"That's different. How do you know, anyway? You're an American."

"Yes. But I once studied a similar disease in New Guinea," I said. "Kuru—caused by virtually the same virus-like agent as that responsible for Mad Cow disease." BSE, scrapie, and kuru are generally believed to be caused by infectious proteins or amyloids, also called "prions."

It wasn't until six years after this Yorkshire dinner—where I ordered chicken—that the first cases of human disease acquired from Mad Cows were reported. A huge public outcry erupted in Britain. Why wasn't more known about this disease? Why weren't people protected? How many people would die and when? How long would it take to know if British beef were safe? Indeed, the outcry almost toppled the British government at the time.

Yet even sitting in that Yorkshire pub, I had seen my friends resisting the notion of these infectious diseases—new, little understood, and having such long incubation periods. More alarmingly, I had heard such arguments before. In New Guinea, too, the people I met wanted to believe they were immune to this strange, elusive infectious agent. Danger is easy to ignore until it hits us in the face.

Kuru had spread in Papua New Guinea (or PNG) through cannibalistic feasts among the Stone Age Fore people and several of their neighbors. (Anthropologists no longer like using the word "tribe," as it implies a political organization with a political leadership, which many loosely organized New Guinea groups don't have. I use the term here not according to its strict anthropological definition, but in a somewhat looser sense.) When a member of the Fore (pronounced For-ay) group died, his or her female loved ones cut up the body, wrapped the pieces in banana leaves or bamboo tubes, steamed these in a fire, and ate the contents. As one woman would later explain to me, "I will now always have part of my mother inside me." As a missionary would soon tell me, "The group's cemeteries are their stomachs." This Stone Age culture, with few material goods, used the body in any way possible. The men and women, I would soon see, hung around their waists on strings—when setting off with bows and arrows to settle disputes with enemy villages—the hacked-off fingers of deceased ancestors. Cannibalism had been widely practiced in New Guinea—both exocannibalism (in which outsiders, such as enemies, are eaten) and endocannibalism (in which members of one's own group are consumed). Yet early in this century, among the Fore, the kuru epidemic began. In this group, it was primarily women, along with their young children of both sexes, who participated in these feasts. Men got priority in consuming whatever pigs and other sources of meat were available.

During the funeral, the brain was given to the closest female relatives, who, with their children, would eat it. This organ, however, contained the highest concentration of the virus-like particle. Yet people saw the brain as a prize—hardly a source of disease—and continued the cannibalism. (In fact, kuru-stricken bodies were prized because the meat was considered leaner.) The mourners, once exposed, generally developed kuru, and at their deaths were eaten, too, further spreading the agent. In some villages, the epidemic eventually killed up to ninety percent of the women and two-thirds of the total population.

Yet in the laboratory, kuru had been transmitted orally to animals only with difficulty, and only to squirrel monkeys, leading scientists to assume that it spread during Fore feasts not necessarily by mouth, but by rubbing hands and fingers infected with the particle into eyes, noses, mucous membranes, sores, and scratched mosquito bites. Hands could remain contaminated for a long time, as there was no soap or tap water.

Once the infection occurs, the disease eventually causes frightening psychiatric and neurological symptoms. Legs wobble uncontrollably and become unsteady. Hands and fingers tremble, and the whole body soon shakes. Patients lose the ability to eat or walk. Speech becomes slurred and thinking slows and disintegrates. Emotions fluctuate rapidly and without cause. Patients burst into fits of laughter at odd moments, leading to the disease being dubbed early on by journalists "the laughing death." Patients eventually grow mute, become demented, and die. In the local language, the term *kuru* means "to tremble."

Dr. D. Carleton Gajdusek of the National Institutes of Health had traveled there to chart the disease in 1956, and eventually discovered that it resulted from infection by a novel pathogen with a long incubation period. He also demonstrated that the same agent caused a rare presenile dementia of worldwide occurrence, Creutzfeldt–Jakob disease. For this work, he was awarded the Nobel Prize in 1976. Many people—from gold prospectors wandering through Papua New Guinea to researchers such as Michael Alpers—had suspected a possible connection between kuru and cannibalistic feasts. Anthropologists Bob Glasse and Shirley Lindenbaum, studying details of the feasts, had helped clarify the connection. Laboratory researchers further probed these diseases in other ways.

Yet, despite these ground-breaking contributions, many aspects of these disorders remained mysterious. These agents have still not been definitively characterized. Their mechanism of replication has yet to be understood. Scientists still don't fully agree whether the clumps of protein that accumulate in brain cells (the so-called "prions") are the cause or the result of the disease, or whether another virus or protein is involved, and if so, what and how. This newly discovered pathogen resists the chemicals and the levels of heat and radiation that destroy all other organisms (hence the Fore failed to inactivate the agent when steaming it during their feasts). The infectious particle has not been found to have DNA or RNA, the genetic material in all other life forms. In addition, the agent can sit quietly inside a body for years before causing symptoms. But how long and why are still unknown.

Gajdusek argued that the number of kuru cases should be decreasing since cannibalism had stopped. Yet the Fore insisted that the disease continued to be widespread. No one knew exactly how many cases remained, or how the disease or its spread were in fact changing. A researcher was needed who had the time and energy to do such a systematic study of individual cases. These questions were important, given how little else was known about these diseases. Specifically—

What was the longest period the infectious particle could take to cause symptoms? Did other factors, including the host's genetic disposition, age at time of exposure, environment, or dose or strain of agent, play any role? How virulent was the pathogen—how often did exposure lead to disease? It was to address these and other related questions that I had journeyed to New Guinea.

I had been interested in going there for several reasons. The connections between biology and culture had long fascinated me. I was born on the Upper West Side of Manhattan, where, as a boy, my parents took me along with my three sisters to parks, the American Museum of Natural History, zoos, and art museums. Biology seemed magical to me. In the park, I used to dig to uncover the roots of trees. On the windowsill of our twelfth floor apartment, I planted seeds from oranges, lemons, and grapefruits that I had eaten. The plants grew less than a foot high. But I noticed that the deeper the pot, the taller the plant. Unfortunately, six- to eight-inch pots were all that I could afford and all that would fit on the sill. One summer, I took a science course and learned how the throat and esophagus worked. I used to draw pictures, as if x-rays, intrigued by how the outside of the body connected to a dark unknown interior.

Our apartment was cramped, the family tense. As an adolescent, my father had inner ear infections, badly treated with multiple surgeries that had cut his facial nerves. One ear remained deaf, his face misshapen, half-paralyzed. My mother's father had died when she was a child. Her mother later forced her to stay home, and not to go away to college. My parents had difficulty giving to their children. "We have the money," my mother would always say, "we just hate to spend it." Tension and frequent fighting erupted. My parents both worked long hours, leaving the house before 7 a.m. and returning over twelve hours later, five days a week. My mother worked on Saturdays, too. They owned a small dress business in the garment district of New York and had to fight daily with factory owners and mafia truckers. My father designed, cut, sold, packed, and shipped the dresses himself. My mother managed the office and sold to walk-in customers. My parents were under a lot of pressure. During summers, my father insisted I work for him as a packing clerk.

I was the only boy and had three sisters. "Get him girls," my older sister would holler every few days and the three of them would come galloping to attack. I'd run as fast as I could to my room and slam the door—which was knocked in four times, and replaced three.

In museums I was free. They were a fantasy world to me. I was drawn most to the Museum of Natural History, and particularly to the longest and tallest hall, "The Totem Pole Room," lined with wooden American Indian carvings (from the Northwest Coast, I later learned). Claws, beaks, and wings rose mysteriously on top of each other into the dark, lofty ceiling, towering over my head. From their soaring heights, eagles stared down, their huge black pupils protruding from concave sockets. For years, these carvings, unlike any animals or people I knew, haunted my nightmares and dreams.

At one end of the hall stretched a long, red and black-painted wooden canoe, in which muscular American Indian figures stood armed with spears, and arrayed in headdresses and native skirts.

At the very back of the hall, a large freight elevator carried me up to a small live-animal section of the museum. Here, colored fish swam in tanks. Snakes curled and rolled, their shed skins flattened and displayed in glass frames above the cages. Their tiny hexagonal translucent scales, each a perfect geometric shape, varied subtly in tone from charcoal to copper.

Before going home, under an old stone arch in front of the Museum building, I would sit on slabs of lava, their surfaces smooth and onyx-like as if from another planet. Beyond stretched Manhattan's cold, rectangular, grey streets.

In 1969, when I was eleven, we moved to Long Island. Dix Hills had been nothing but fields until only a few months before. Our whole Levittown development had been a rhubarb farm. The top soil had been bulldozed off for sale, leaving thick beige clay. We visited the model house. My parents decided not to pay the extra two thousand dollars to be in the woods. They signed the papers, and Levitt built a house for us, too. The first time we visited it, the road itself had just been bulldozed. I got out of the car, and stepped onto a vast treeless stretch that was going to be our front yard. I sunk down into mud up to my calf. Years later I would learn the hard way that the clay barely supported grass—except for crab grass and weeds—without seasonal rounds of Miracle Gro. Two months after our first visit, I left behind my friends in the city, and began my new life.

There was no town, history, or culture. Everything had been built at most only a few years before. On weekends, everyone hung out in malls. Shiny chrome and glass, evanescent fashions, and big, blown-dry hairdos filled these long glass cages. I yearned for something more permanent and universal. The only culture here was the library. There, one day, I found brochures for a summer course at the Cold Spring Harbor Laboratories, located twelve miles away and run by James Watson, who with Francis Crick had received the Nobel Prize a few years before for discovering the structure of DNA. I decided to go. I would bicycle there and back, if I could get there no other way.

I took one course in geology and one in marine biology. In the first course we drove around Long Island in a big yellow school bus examining geological formations left by glaciers twelve thousand years before, learning how these ice sheets worked. Long Island was flat as a pancake. The tallest hill, which we walked up, was only four hundred and fifty feet high. The whole south fork was originally a series of islands linked later by outwash plain. Each island was a huge deposit of rubble left by the ice as it departed into the sea.

Only three rocks on Long Island are bigger than a person—having been carried there by the glacier. We saw all three. It was the first time I ever went to the Hamptons—we toured exclusive golf clubs to see the formations.

In the marine biology course, we caught different kinds of crabs as they scuttled across beaches and dunes. We examined the creatures under microscopes, and mapped out the habitats of different species. We watched horseshoe crabs—one of the oldest surviving species on earth—flip themselves over on the beach after crawling up to lay eggs. If they got stuck, they would die. With their foot-long spike in front and curved brown helmets, these creatures seemed primordial—left over from some long-lost world.

When I learned to drive, I found another escape. My family had one car and all of us had to share it. But I loved cruising down the Long Island Expressway—the night sky above, darkened trees on either side, the night road stretching indefinitely, it seemed, before me. Yet the road could never take me far enough away from my pressure-cooker of a household.

Most of the bright kids in junior and senior high school wanted to be doctors. I didn't yet know. Science interested me. But I didn't think much of most of the doctors and dentists in my neighborhood. I felt a need to choose a career—whether medicine or not—on my own terms, and to find out more about it first.

But after seeing my parents work, I knew I could never do business for its own sake. I am also descended from a long line of rabbis—my grandfather, two great-grandfathers, and a few great-great-grandfathers.

My father rejected this tradition and pursued business instead. But somehow a desire to understand ultimate things was passed on to me. I thirsted for knowledge of essentials and truth about human beings, myself, and the world.

When I left for college, I didn't know what I'd do. Scientific research seemed exciting, but intimidated me. Who was I, just a kid, to presume to do it? But I was looking for something pure, and unchanging, and I was eager to leave suburbia and materialism behind.

In college, I was drawn to biology, but also, to my surprise, to literature and intellectual history, specifically Rousseau and Nietzsche—to questions of what was universal and hence biologically based about human beings, and what was specific to different times and places. How did culture, as opposed to nature, mold men's and women's views and experiences of the world and themselves? In college, academic disciplines seemed to perceive and study human nature differently—sociologists looked at group interactions, psychologists at individual cognition, biologists at human physiology, literary critics at people's writings, historians at past events. Each field seemed to feel it uniquely possessed the most vital set of explanations. But I kept thinking of the parable of the blind men and the elephant. Each discipline understood only part. None seemed to grasp the whole. I wanted to understand underlying human nature—men and women at their roots.

Practically speaking, I thought of two possible ways to pursue these interests—through the humanities and writing on the one hand, and through medicine and psychiatry on the other. Yet I hesitated becoming an academic or a writer, removed from the world. I liked interacting with people and seeing how others lived. I didn't want, for example, just to write for writing's sake. Medicine and psychiatry appealed to me as ways to study and understand our species. But these fields could be abstract and cold. They also involved caring for patients and being close to illness and death all of the time—none of which I had yet done. I was wary that these fields could also quickly become routine and mundane. To learn more about medicine, I decided to take a summer job at the National Institutes of Health—one of the world's premier medical research institutions. I didn't know if medical research would interest me, but I thought I would try it. That is how I first heard of kuru.

Fort Detrick

"That's the Anthrax Building," Dr. Elsa Rojas told me, pointing as we drove by a six-story red brick building surrounded by a quiet, neatly mowed lawn. Rough grey cinderblocks sealed all the windows and the wide doorway. No cars or people were anywhere in sight. I had just arrived at the National Institutes of Health (or NIH) labs at Fort Detrick, Maryland, to work.

"What's inside?" I asked.

"Anthrax. It escaped forty years ago. The building has been sealed up ever since." Neither she nor anyone else ever mentioned it again, though we would all drive by it each day. The deserted structure seemed somehow haunted.

Gajdusek's labs at the main NIH campus in Bethesda and here at Fort Detrick were the major ones in the world conducting research on kuru, Creutzfeldt–Jakob disease (or CJD), and other infectious protein diseases. CJD, a presenile dementia caused by infectious proteins, kills one person per million per year throughout the world. Over two hundred people die of CJD in the United States every year; several years ago, it claimed the choreographer George Balanchine. At Balanchine's death, scientists wondered whether the disease would now be renamed "Balanchine's disease," just as another neurological disorder—amyotrophic lateral sclerosis—became "Lou Gehrig's disease" after killing this baseball hero. Yet a patient, to have a disease named after him, seems to require a certain wide, popular celebrity status that Balanchine apparently lacked.

CJD can occur spontaneously without any known exposure. The kuru epidemic may have resulted from a Fore member contracting CJD.

On his or her death, close kinsmen contaminated by the cannibalistic ritual became infected, developed kuru, and were consumed as well, widening the epidemic. These agents, including other, even rarer neurological diseases, such as Gerstmann–Sträussler–Scheinker syndrome and fatal familial insomnia in humans, and transmissible mink encephalopathy are cumulatively known as transmissible spongiform encephalopathies.

Most recently, CJD is also the illness that people have gotten from Mad Cows. Almost all that is known about these agents in humans has been learned from CJD and kuru. New Guinea contained the world's largest concentration of human cases of infectious protein disease.

Leading scientists have argued that consumers of beef may be in danger in the United States as well as in Britain, as inspection standards have recently slipped in this country, with the introduction of a cheaper, less effective means of grinding up and processing carcasses for cattle feed. In fact, some scientists have suggested that the rise in Alzheimer's disease in the United States over the past few years may represent misdiagnosed cases of CJD, often indistinguishable by clinical or standard laboratory tests. No case of Mad Cow disease has yet been proved in the United States, but many cases, traced to cattle or feed imported from Britain, have been identified on continental Europe. In Britain, problems remain despite the slaughter of millions of cattle. The agent can persist in soil for decades. After scrapie breaks out among sheep, a farmer routinely will kill the entire flock to prevent wider contagion. Unfortunately, a new flock introduced into the same pasture even decades later will develop the disease as well. How the infectious particle survives and thus spreads remains unknown.

This disease is important to understand, as it may shed light on other disorders as well, including Alzheimer's disease, multiple sclerosis, and Parkinson's, which affect millions of patients and impact the lives of their loved ones, too. Specifically, in both CJD and kuru, the central nervous system degenerates and symptoms including dementia and poor coordination result. Indeed, the late doctor and writer Lewis Thomas said that if he were a young man just starting his career, he would focus on what he considered the most exciting new area in medicine—these strange contagious agents.

Yet these diseases remain elusive and mysterious, because of the difficulty of conducting research on them, stemming in part from their long incubation periods.

However, investigation of such epidemics has become increasingly important as new technologies and more frequent travel link populations more closely. Diseases can now more readily cross the boundaries of their prior natural habitats.

Here in Fort Detrick, a maximum security military base, biological warfare experiments had been conducted in the past. Tall white wooden guard towers with small shuttered windows stood at the gate. High metal fences topped with thick coils of barbed wire encircled the fort. Beyond the wire rose the blue Cachuctin Mountains, hiding other military bases—also carefully patrolled by uniformed guards—and the presidential retreat at Camp David. I would work here at Fort Detrick and also in Bethesda.

Dr. Rojas escorted me now to Building 84, and entered a code on a keypad at the doorway. Inside, along the wall, shelves upon shelves supported lidded circular glass tanks—the size of ladies' hat boxes. In each jar, a human brain floated, infected, she told me, with kuru or CJD. The organs swam in pale green fluid, suspended on their sides or facing forward, sunk to the bottom or pressed against the glass, as if staring out. Scotch-taped to each glass drum was a white index card labeled with letters and numbers in red magic marker. The pinkish-grey brains lay captured, as if primeval creatures from the deep. Convoluted twirls, resembling coral, covered each. I had difficulty imagining that each of these mushy blobs once produced dreams, laughter, tears, and song. From each bulk, magic once flowed, and had been taken away by a tiny protein.

"These are our P4 facilities," Dr. Rojas told me, showing me another room. In the middle of a special cinderblock chamber stood complicated apparatuses—large Lucite boxes with tubes leading in and out, and thick plastic gloves pointing inward—maximum protection to prevent the release of dangerous particles. My hands, entering through two portholes into the gloves, would never have to touch the deadly agents themselves. The rooms also came equipped with reverse suction. If anything went wrong—any dangerous molecules escaped—all air would be sucked out of the room. The lethal material wouldn't be let loose. (But I wondered if any people left behind would then suffocate and die.)

"We're required to have these rooms here," she told me as I followed her back out, "but we rarely use them. They're too cumbersome to get anything done in. And it's not completely clear we need them. So we usually just work as carefully as we can in our more standard labs."

"Also, we're supposed to wear these outfits," she said, showing me a box on the floor. I pulled one out—a special white plastic jump suit with a zipup front. But, I would soon see, we almost never wore the uniforms. They were poorly ventilated, and made us sweat. "We've probably all become exposed to these viruses," she said. Yet she and other researchers continued to work, undeterred.

A few minutes later, a neurosurgeon stopped by the lab and introduced himself. "You're in luck," he told me. "You're going to get to see a live brain today." I was excited, never having seen one before.

Half an hour afterward, he shepherded us into a small room down the hall where he had strapped down a monkey. The tiny animal had been fed part of a human kuru-infected brain to see if the disease agent could be transmitted orally. The dark brown furry animal glanced anxiously with big eyes over at the neurosurgeon and then me. The neurosurgeon anesthetized the small creature through an oxygen mask and then held up a small T-shaped tool resembling a corkscrew wine opener. "This is the same instrument the Incas used for neurosurgery," he said. He propped it against the side of the monkey's head and turned it, drilling a small hole above the ear. "Here, have a peek." I felt terrible for the monkey, but stretched to peer into the dime-sized opening the surgeon had bored through the skull. Shiny red arteries and bluish veins covered a pulsating shiny pink membrane. I had never known that the brain beat—it appeared almost like an animal with a life of its own. The researcher poked a needle into the brain to excise a tiny sample. Here indeed lay secrets of the disease, of the nervous system, and of nature itself.

Back in Dr. Rojas' lab, over the next few days I started helping with experiments designed to determine whether the infectious particle in fact contained any genetic material—DNA or RNA. I mixed masses of kuru-infected brains with various chemicals to form a clearish pink extract that had to be stirred for several days. I poured the mixture into a tall Ehrlenmyer flask—with a round bottom and a tall narrowed neck that impeded fumes from escaping—dropped in an oblong white plastic magnet, and placed the container on an electric plate that made the magnet spin around to stir the soup of brains. I set the whole apparatus in a tall, glass-fronted refrigerator where the magnet would turn for several days.

In the meantime, there was little to do.

As my experiment sat in the fridge, I started visiting researchers upstairs at the NIH who studied the anthropology and epidemiology of these diseases—how, when, and to whom these illnesses spread. This more human research, I quickly realized, could reveal other mysteries—just as important—about these diseases, and illustrate how culture and illness interact.

Two weeks after starting in the lab, I was told by Dr. Rojas that for the next experiment we would each kill one hundred kuru-infected mice and remove their brains. She showed me how to pick up one white rodent at a time by its tail from a huge cardboard box. I followed her example. The mice scurried as best they could away from my gloved hand as if aware of their fates. I finally grabbed one animal by its tail, and lifted him onto the taped-down rough blue paper in front of me. The mouse ran forward as fast as he could. As instructed, I pressed down on the back of his neck with the edge of my scissors, and yanked his tail until I heard a snap. I had now broken his spinal cord. I sliced open the back of his head with a scalpel, peeled back his white fur, cracked his skull, and plucked out his brain with a tweezer.

Dr. Rojas didn't seem to mind the work, but it turned my stomach. I don't remember how many mice I managed to kill that day, but it was far less than my quota of one hundred. These lab animals enabled important insights to be made into the nature of these infectious agents and could lead to the saving of countless human lives. Many people kill animals bothers many of us—for food, for science, for clothing, or for sport. Some of these purposes are more sanctioned than others (not everyone advocates vegetarianism). Yet clearly the slaughter of animals bothers many of us—the reasons have to be good. In this lab they were. But I myself couldn't do it. I started to stay upstairs in the anthropology section more, and soon arranged to work there full-time.

Eventually, this experiment and others failed to find any DNA or RNA, making these agents the only known life forms that apparently lacked either form of genetic information. Indeed, most recently, Gajdusek, Stanley Prusiner, and many others believe the infectious agent is a normal protein in the cell, a molecule present in everyone's brain, that gets twisted in the wrong direction. The deviant protein either flips on its own or enters the body from the outside. Once present, the protein acts as a crystal, inducing other, normal copies of the protein to reconfigure as well, precipitating clumps of protein that clog up the cell. The process resembles cloud seeding with crystals of silver iodide that serve as nucleants around which raindrops form. Similarly, Kurt Vonnegut, whose brother, an MIT physical chemist, had worked on cloud seeding, described in his novel *Ice-Nine* a differently shaped ice crystal that can solidify the world's water. So, too, a malformed protein could nucleate conformational change in normal protein that would then polymerize as precipitated crystals in the brain. These abnormal molecules, acting as nucleants, get further duplicated. Eventually, millions get made, destroying the cell, and escaping to attack and kill other cells as well. Large holes soon pockmark the brain, making it look under a microscope like sponge or Swiss cheese (leading to the disease's classification as a "spongiform" encephalopathy). As the protein belongs to the brain and is not foreign—simply misshapen—the body does not attack it by mounting an inflammatory response with fever and pus, as in other infections. The body does not know that it has been invaded.

Many questions remain unanswered. How does this abnormally folded protein manage to make these similarly misshapen copies of itself? Why can the pathogen take so long before beginning to have an effect, and what is it doing in the meantime? Does the agent become activated spontaneously or as part of a process—either external (e.g., prompted by an outside environmental trigger) or internal? As these questions remained unanswered in laboratories, data from natural laboratories—from social settings where the disease has appeared—were crucial. The disease among humans—rare except in New Guinea—needed close investigation.

As a newly discovered life form, even its name is hotly debated. I was asked to help to decide what to call it—whether an "unconventional slow virus," an "infectious protein," an "infectious amyloid," or a "viron" or "virin." Stanley Prusiner, who in 1997 won the Nobel Prize for his work pursuing the mysterious agent, coined the term, "prion," derived from the word protein. At least until this second Nobel Prize was announced the field has remained polarized in the nomenclature it employs—"infectious proteins" or "infectious amyloids" on the one hand, or "prions" on the other. Despite his contributions, Prusiner has earned some critics who have suggested that his term derives in part from his own name.

I would have nominated another term, more descriptive of the agent's unique characteristics—a crystal virus, or crystal protein. Yet I suspect Prusiner's Nobel Prize will significantly strengthen acceptance and use of his term. The label is shorter and simpler—demonstrating the importance of language in science. Elegance of phrase goes a long way in written and verbal communicatioin both formally and informally among scientists and others.

One day at the lab, Carleton Gajdusek called, having just returned home from New Guinea. He asked for someone to come over and help open the mail. An anthropologist asked me if I wanted to go, and I leapt at the chance.

Gajdusek's house was crammed with artifacts from New Guinea and other far-off lands. Bookshelves and cases brimmed with literature, art books, and anthropological treatises on cultures and civilizations from every corner of the globe. The smell of old and musty books, and of smoke-scented wooden carvings from jungles around the world filtered through the rooms. "Ideas are cheap," he said, sitting down. "What makes great science is collecting data in a systematic, plodding, and routine way. It is the repetitive doing of a physical task that reveals discrepancies and exceptions and thus insights, and can lead to world-class science or art."

Intellectually, Carleton has been the most remarkable man I've ever met. I have been fortunate to have spoken in my lifetime with many accomplished individuals—U.S. presidents, vice-presidents, MacArthur "genius" awardees, Pulitzer prize winners, and other Nobel laureates—but Carleton remains by far the most impressive. The peculiar sensation I had when first meeting him I have found described best in *The Autobiography of Alice B. Toklas*. A little bell went off in Alice's head when she first met Gertrude Stein that said, "genius." This is what I felt when I first met Carleton, and which I have felt only two or three other times in my life (most notably with Harold Bloom and Norman Mailer). Indeed, Harvard Medical School, to celebrate its 200th anniversary a few years ago, decided for the first time to award three honorary degrees, and choosing among the scores of Nobel laureates who had passed through the institution over the years, gave the first prize to him. Gajdusek had also brought to the United States dozens of New Guinean and Micronesian boys and girls whom he sent to elementary and high school and college at his own expense. When I met them, the New Guineans began to teach me Pidgin English.

(Sadly, Gajdusek was recently charged with having sexual contact with one of the 17-year-old Micronesians. In a plea bargain, Gajdusek began in April 1997 serving one year in jail. The case raises complex issues. In traditional Melanesia, sexual contact between adults and children is sanctioned, indeed the norm. Yet though cultural practices and taboos are relative—what one culture encourages, another may prohibit—violating a taboo in any one culture can lead to punishment. My own contact with Gajdusek has been an important part of my education, intellectually and scientifically. I was disturbed to learn through the court case of the activities in which he reportedly engaged privately. Yet though I initially questioned what I had learned from him, in the end the lessons he taught concerning science, medicine, epidemiology, and anthropology continue to be valid and inspiring to me.)

No matter what I said to Carleton, what observation I'd make, he'd see something further, make some added insight, take apart my understanding, and turn it on its head. At one point he said he wanted four old file cabinets painted. I noticed he had four different cans of paint—yellow, orange, blue, and red. "Which color do you want?" I asked.

"Just pick one."

"How about using all four?" I suggested, thinking it would make the cabinets more colorful and fun.

"Great idea," he said immediately. "That way I can say to one of the kids, 'put this in the third drawer down in the yellow cabinet, or the top drawer in the blue one.'" He instantly saw a useful and practical implication for what was for me merely aesthetic whim.

Gajdusek had been instrumental in all aspects of research on these diseases. He could discourse readily on these crystalline particles, having studied with Linus Pauling at Cal Tech, an expert in physics and chemistry who had won two Nobel Prizes. Gajdusek also had a rare understanding and appreciation of the concrete physicality of things—the materiality and very substances of which they are made. (He once told me that if he had to do it again, he might have been a sculptor.) This feeling helped him, it seemed to me, in conceptualizing genes, cells, and eventually infectious proteins.

I soon started to work under Dr. Gajdusek himself, cataloguing and analyzing records and films of primitive peoples in New Guinea. I watched and catalogued reels of anthropological films, unpacked boxes of artifacts and jars of native medicines, and studied diseases and populations in New Guinea and on isolated South Pacific atolls. I also began to talk to him about the history and philosophy of science, which I was studying in college, and about primitive and modern beliefs about medicine.

I was impressed by the researchers I met at the lab. At the end of the summer, when I returned to college, I stayed in touch with them.

Back in cloistered Ivy League classrooms I felt bored and restless. Around the large wooden table of a humanities seminar, discussions of abstract ideas about a novel seemed like mental masturbation. Others seemed not to mind it, and in fact to need it more than I did. To experience and engage with the world appealed to me more. That semester, I also read Conrad's *Lord Jim* and Malraux's memoirs. I began to dream of adventures in far-off climes—of which there weren't many left—and of writing, traveling, and exploring the world. I didn't like secluding myself in my study. One midnight, as I stumbled back to my dorm from the pub, I noticed the lights still on in one of my professor's offices. She was busy at work. A few months later she was denied tenure. That was not the life I wanted.

But I was unsure what the future held. I had heroes: among them Conrad, Malraux, and Hemingway. But I was merely a young Jewish boy from Long Island. What was I going to be able to do now at the end of the 20th century when there were no longer wars against fascism, and the world was more explored than it had ever been? I had never even been out of the country.

I wasn't sure how to choose a career—how daring or conservative to be. I also saw a tension between pursuing science or more humanistic studies, and wasn't sure on which to embark. One day during my senior year I had lunch with a literature professor. I mentioned that I was considering medicine as a possible career. "I am interested in the body, too," he said, "but do you really want to be dealing with *sick* people?" He sat in his tweed jacket, his fingers hanging over his cellophane-wrapped food. Yes, I then realized, I wanted to deal with something more than books alone.

I asked my parents what they thought I should do. My mother liked the idea of me being a doctor, and didn't encourage me to do anything else. Yet I remained unsure. "Okay, well ask your cousins who are lawyers and doctors," she finally said, "and see what they think." At Thanksgiving my senior year, my cousin who was a lawyer said don't go into law. My cousin who was a doctor said don't go into medicine.

I returned to college after Thanksgiving break, arriving there late at night midst dark pine trees and a full moon. A quiet calm hung over the campus. I was glad to be back. The university stood for something pure: Science, Truth. "Here," a carved inscription over a doorway that I passed read, "we were taught by men and Gothic towers . . . of unseen things that never die." I wanted to study men and women in their natural state, removed from social claptrap.

Over the next few months I applied to medical school, but knew that I wanted to do something else first. Over Christmas, I completed and sent off medical school applications, then saw the film *Lawrence of Arabia*. I was swept up by the tale of a young man in his early 20s, recently graduated from college, adventuring in a faraway land. The message: one can set one's mind to do things and accomplish them. I yearned even more to travel to some foreign part of the world. Perhaps I was merely being a romantic idealist. But I was drawn to adventure.

I had arranged to complete some of the projects from the lab while at school, and when I finished, over Christmas break, I called Gajdusek up. He asked me my plans for after graduation.

"Medical school," I said, "but I thought I would take a year off first. I applied for a Fulbright to study in Switzerland for a year."

"Why don't you do something interesting?" he said.

"Like what?" I asked, surprised he didn't think my plans already were.

"Why don't you go to New Guinea?"

"All right," I said, not sure exactly what kind of commitment either of us was making.

The island of New Guinea remained as unexplored as Africa was in the nineteenth century. Only recently, Carleton and his colleague, Michael Alpers, had accidentally discovered a new volcano there. Until the 1930s, the entire highlands had been unentered from the outside, making it the last unknown territory on earth. The first Western explorers—two Portuguese, Antonio Abreu and Francesco Serram— saw the coast of New Guinea in 1512, but didn't even give it a name. Don Jorge de Menese, also Portuguese, accidentally ran into the island in 1526 and named it "Os Papuas"—"frizzy haired" in Moluccan. The name stuck. A Spanish explorer, Ynigo Ortiz de Retes, seeking a better route from Indonesia to South America, later called the island "New Guinea," thinking the natives resembled those of Africa's Guinea coast. Half a century passed before the next European arrived.

Over the following two hundred years, Spanish, Dutch, British, and French ships came and went along the coast, though not until well in the nineteenth century did any colonization begin. In 1828, the Dutch, owning the neighboring East Indies and the Mollucans, annexed the western half of the island of New Guinea, though then did little with it. Still, nothing else happened until the 1880s, when the Germans under Bismarck, seeking to become a world power, set their eyes on the Pacific, and on the northern half of the remaining part of the island. England's colonies of Australia and New Zealand begged the British crown to claim the remaining southern half, which Great Britain belatedly and reluctantly did. Still, exploration of the island moved in slowly. The unnavigable, malaria-infested rivers impeded further probing of the island. Outsiders contacted only the coasts. The center remained wholly unexplored. Not until the 1930s did the discovery of gold impel a handful of prospectors to venture in. To their surprise, they found the interior inhabited. But even the promise of lucre failed to overcome barriers of environment and climate. Only with airdrops and two-way radio was the island slightly more surveyed. The outside world still remained indifferent until World War II, when the island became an important battleground. The Japanese, pushing to control the Pacific, coveted Australia as an ultimate prize and invaded New Guinea. Bitter fighting raged to regain control of the island. For the first time, airplanes flew over many regions. In darting over the steep ranges, some crashed.

After the war, systematic exploration finally began of the eastern half of the island, which was found to contain three million inhabitants, still living in the Stone Age. Groups, tucked in steep valleys and surging mountains, had remained markedly isolated from both each other and the outside world for millennia. On the island of New Guinea, over seven hundred and fifty separate languages are still spoken—one-half to one-third of all the languages ever spoken in the history of the entire world. These tongues are not dialects with differences akin to those between, for example, French and Italian, but vary from one another as markedly as Hungarian from Hawaiian. Cultural isolation couldn't be more extreme. The Fore are only one of these seven hundred and fifty groups.

Anthropologists such as Bronislaw Malinowski and Margaret Mead had come to various parts of New Guinea and its islands since the 1920s, and each had examined a different set of people. These researchers explored an assortment of themes, illuminating the various ways in which cultures approach warfare, power, spirituality, gender, sexuality, and aesthetics. New Guinea has served as a vast lens and mirror for understanding human culture and psyche. Over the years, a few additional anthropologists had ventured into the Eastern Highlands, including Ronald and Catherine Berndt, Bob Glasse, and Shirley Lindenbaum. But crucial, otherwise unattainable data remained there— soon to be lost.

As kuru constituted the vast majority of the world's cases of transmitted infectious protein disease, it was also important to find out as much as possible about this disease. But the task would not be easy.

Recently, several puzzles, medical and anthropological, about the illness had been appearing. While the natives argued that the disease can be cured through counter-sorcery, Western scientists viewed the illness as untreatable. Did the Fore possess a remedy? If not, how did they support their claim despite evidence to the contrary? As the natives also insisted that the number of cases remained the same, one of my jobs would be to determine who had kuru by Western criteria and who did not, in order to understand whether and how the epidemic was changing.

The Fore and their neighbors still believed that sorcery caused the illness. As a result, the Fore had continued cannibalism (in respect for their dead relatives)—even in the face of mounting deaths. As more tribal members died, more were eaten. In fact, cannibalism had begun to decline not as a result of kuru, but from colonial government patrols and missionaries, arriving for the first time in the decades after World War II, pressuring the natives to end the practice, and arresting any violators. Occasional cannibalistic feasts are rumored still to occur in secret. Had Caucasians not entered the area when they did, the Fore may have vanished.

Yet the transmission of the infectious agent through cannibalism had not actually been proven. In fact, an anthropologist, W. Arens, had published a book in 1979 entitled *The Man-Eating Myth*, arguing that cannibalism had never existed anywhere in the world, and had merely been made up by Westerners for prurient reasons. The nation's leading anthropological journals and *The New York Times* reported on this new idea despite it being purely speculative and unfounded. This "revisionist" theory insisted that proof of cannibalism exists, at best, only as written records that are merely "texts" and do not refer to anything in reality. Hence, another of my goals would be to document the lingering memory of the nature of these feasts, adding these data to the record. No one born after cannibalism stopped had been found to develop kuru, but it would be important to show that this was still the case—particularly now with more recent patients.

Since the length of time from the most recent cases of cannibalism was increasing, it might also be possible to determine how long the incubation periods could be—how slow these "slow viruses" (as they were then called) were. How long *could* the infectious particle take to affect someone after exposure? Previously, it had been impossible to trace cases back to specific feasts because so many feasts had been held in a very short period of time. Yet current cases might be the results of the last feasts held. These recent patients might have attended only one or two feasts in their lifetime—as children—not dozens or hundreds throughout the span of their lives as had been the case with patients earlier. Thus, it might be possible to pinpoint exactly when infection had occurred and exactly how long the agent took to begin its attack.

Recently in some villages, after no cases at all for several years, clusters of individuals appeared to get sick suddenly—after decades of being healthy—and die. It was not clear whether these were new outbreaks, and if so, what caused them. Was the infectious agent indeed still lingering somehow in the environment? Or were these clusters the result of the last feast or feasts? These individuals would together then have been infected decades before and been completely healthy until now, when the slowly replicating agent attained levels sufficient to kill. The idea sounded absurd—something out of science fiction. But the discovery of identical incubation periods of possibly thirty years in two or more people would suggest that a very specific, far from random process was occurring inside the brain. Moreover, a second, environmental "trigger" would then be far less likely. These specific feasts, if identified and shown to have thus transmitted the disease, would also serve as documented cases of culturally sanctioned cannibalism, providing even further evidence that cannibalism did in fact occur. Moreover, if such episodes of transmission were found, these decades-long incubation periods would then be measurable and confirmed for the first time in particular individuals and would be the longest ever documented. These discoveries could help alter how medical science thought about these and other disorders. Most infectious diseases—such as the flu—incubate merely for days, or even hours. Other adult ailments, too, might perhaps then be the result of virus-like particles acquired decades before but not producing the usual immune signs indicating infection.

"If you're able to find such clusters," Carleton told me, "it would be an important contribution to our understanding of these diseases." It would be vital to collect data on all current and recent cases—many of which would not be seen or documented by Westerners otherwise.

As the world was shrinking, becoming increasingly homogeneous as a global economy, with the same products available in every country, I was excited about the prospect of exploring first-hand a land as different from my own as possible. There was perhaps no better way to understand what was universal about man (by which I will mean human beings—both men and women) and what was specific to different times and places. I had studied these issues through works of history, sociology, anthropology, and literature. But I now wanted to look for myself. I would approach these questions by exploring how a different culture viewed death, brain disease, and epidemic. Western medicine looks at illness, the brain, and the human body in specific ways—scientifically. Nothing could address questions about culture and psyche better than comparisons with the Stone Age and New Guinea—the last region contacted by the outside world and the home of the largest concentration of primitive people on earth. What was life like in the Stone Age, untouched by civilization (using this term loosely)? This land could offer myriad similarities and contrasts for understanding our own society—from how people dressed, to how they viewed life and mortality, and what they dreamt about.

The Stone Age and the New Guinea rain forests were also now disappearing. Coffee plantations and foreign lumber and strip-mining companies were rapidly denuding mountains and valleys. I would be exploring a universe and a time that were soon to vanish. In addition, the governments of New Guinea and the surrounding islands were now wary of anthropologists coming in to "study the natives," and rarely issued visas to social scientists. Reports on the primitiveness of New Guinea did little to entice foreign investment. Yet I would still be able to enter this unique area of the world, this natural laboratory, as a medical researcher, and I would thus have one of the last chances to explore cultures there. Moreover, it would be important to observe how the Stone Age interconnected with the Space Age and the information superhighway—how individuals leaped across thousands of years of development in a single lifetime. Many had been born in the Stone Age before Westerners had arrived. How did they put the two eras together? What was easy and what was difficult about the integration? How did the two cultures—theirs and ours—view and interact with one another? What was different and what, due to man's underlying biological nature, was the same? Brazil had primitive peoples, too, but they had had contact and colonial pressure for over five hundred years. PNG had not. This trip to New Guinea would thus provide unique glimpses of radical shifts in societies and individual lives.

I didn't yet think much about the obstacles to doing this research—having to communicate in Pidgin English, facing malarial swamps, parasites in the water, landslides, tribal wars, and other natural and human dangers. I felt a little scared, knowing no one there. I had never seen dying patients and was afraid of dangerous infectious diseases. But there were important questions to answer here. Also, I figured that afterward I'd probably stay in large American cities for years. This, then, would be my last chance to immerse myself in a completely different place.

I spoke with Carleton further about going.

I applied for a visa from the New Guinea embassy—which told me the processing could take months. In the meantime, I edited a catalogue of Carleton's films on kuru and primitive cultures.

"What should I do there exactly?" I asked Carleton one day.

"Mike Alpers, the Director of the Institute of Medical Research in Goroka, will have instructions for you when you arrive."

"Any survival tips for working there?"

"Take as few men as possible on patrol with you," he said. "They'll all want to be paid and fed. Outsiders will also cause trouble in your host village." It seemed odd as his only piece of advice—not more conceptual or intellectual.

"Do I need vaccinations?"

"Yes. And you should be on antimalarial medication. Take Fansidar. You'll probably go into Fansidar-resistant areas, however. You can take chloroquine, too. But then you'll probably be going into chloroquine-resistant areas as well. So it's best probably just to stay on Fansidar, and when you're on the coast, double your dose."

"Will that work, though?"

"It should."

Still, I was scared. The year before, the NIH had sent a young female researcher to PNG. After one week, she had a nervous breakdown, and had to be flown home.

In the final weeks before my departure, I scurried about finishing projects and preparing for my trip and for being away from civilization for several long months. The week before Christmas, two days before the embassy knew my flight was to leave, I received my visa.

The evening before I departed, I finished last-minute Christmas shopping in New York. I would give the presents to my family when they saw me off at the airport the following morning. Shoppers shoved into store elevators. Crowds pushed. I yearned to escape this crass materialism for something more innocent and pure. In one store, midst mobs and elevator bells, five children, all but ignored, sang "Silent Night," their voices lost in the trample.

That night I had a drink with friends at a midtown Hyatt hotel. Most, already tired of their jobs in New York, wanted a change. I was about to have one.

Goroka!

The next morning, I boarded a plane. The only direct flight from the United States to New Guinea departed once a week from Honolulu on Air Niugini—Pidgin English for the name of the country. I would fly first to California (where I had never before been), and then to Hawaii.

High above the clouds, I finally sat back and relaxed. In Los Angeles I visited college friends briefly and then left the continental U.S. Back in the air, clouds hung over the small fragment of California as if thin paint on the limitless blue. In less than a year, I'd be entering medical school. But now, I was free.

I arrived in Hawaii at night. In the morning, I strolled to the beach and faced the Pacific for the first time. It was New Year's morning, 1981. The ocean stretched on, fusing with the sky in diffuse frothy light. The ocean seemed far softer than the Atlantic—the only other ocean I knew. On shore, rich, fragrant air filtered through lush trees. Cypress, birches, and palms surrounded me. Here was paradise, eternal summer, a dream. I had forgotten how beautiful nature could be. I now understood the attraction Pacific islands had for Gauguin, Robert Louis Stevenson, Melville, and Captain Cook.

The beach was deserted except for a woman sitting in jeans and a work shirt on a square blue blanket. She caught my eye and we nodded. I noticed she was playing cards—tarot cards on closer inspection. "Do you want me to tell you your fortune?" she called out. I strolled over. "Pick a card and turn it over. It will be your fortune." My hand seemed to stop over one. I flipped it over. She was astonished. "You chose the Ace of Cups!" she exclaimed. "You have a very full year ahead of you."

The next day, I boarded a plane for Papua New Guinea. I felt a sudden sadness, about to leave the United States for almost a year—the longest I'd ever been away. Before entering the airport terminal, I took one last long glance around me—at the West. The branches of trees swayed quietly around me. I picked up my bags and walked toward the building. In the tinted glass doors I glimpsed my reflection, standing by myself in jeans and a workshirt that had big breast pockets. I was lugging two big suitcases and a knapsack—a traveler, bound for the unknown.

I was curious to see who else would be journeying to this remote island—I knew no one else there.

Only fourteen people boarded the huge 747. Rows and rows remained wholly empty.

I took three seats. As I buckled my seat belt, I still felt apprehensive—that I should know more about Papua New Guinea, and what I was getting myself into. I also feared becoming lonely or homesick. I had a long list of questions based on readings in psychology, anthropology, and literature: how do dreams and myths differ between cultures? Is man by nature alienated from his society? But these issues now seemed distant and abstract. I looked out the window.

"Is anyone sitting here?" a plain faced man with white hair suddenly asked me, standing in the aisle, waving his hand.

"No," I said hesitantly. He slid along the cushioned seat beside me. Curious sounds came out of his mouth. He caught my uncertainty. "I lip read," he explained mumbling. "I was born deaf."

It seemed an overly personal thing to say to a stranger. "Yes," he continued, sensing my surprise. "Before I was 8, my parents thought I was insane. I got used to getting spanked. But living on a farm, I watched my father doing chores, and I began to do them, too. He realized that an insane person could not do these jobs correctly. I was taken for tests and found to be deaf. It took me to age 16 to learn how to lip read. Eventually, I trained as a carpenter. Now, every year for three weeks, I travel to New Guinea to work at a mission and teach carpentry."

A tall black man walked up the aisle and stopped beside us. "Hey, how are you doing?" he said to my neighbor and me. "I'm Walter," he said. He was in his early 20s. He asked me what I did, and what I would be doing in New Guinea.

"Medical research," I said. "How about you?"

"I'll be doing some preachin' there. I'm a quarterback for the Los Angeles Rams. I've been born again and now spend a few weeks each year spreading the word."

"I see." I was wary of missionaries, and thought of them as merely the justification for colonial powers like Spain and Portugal to have taken over South America a couple of centuries ago. Political and economic interests always seemed at play.

"What's your religious background?" Walter asked me.

"Jewish."

"Oh," he said, as if stopped short in his tracks. "That's cool."

Just then an elderly woman with frizzy white hair approached us in the aisle. She stopped to hold onto the top of the seat. Two younger women followed her closely, watching her progress. "Hey, how 'ya doin'?" Walter turned and called out to her. His forwardness amazed me, but the old woman's eyes lit up.

"I'm just fine, thanks," she answered gladly. "How are you?"

"Terrific. Looks like you're off for a big trip," Walter said.

"I certainly am. I've always wanted to go to New Guinea. If I don't go now, I don't know when I would."

"You're on vacation?"

"Oh sure," she laughed. "I stopped working years ago. I was a pediatrician."

"Bob here is a medical student," Walter said, introducing me.

I stood up and shook her hand. Hers was frail but firm. Her name was Betty. "When I was in medical school," she told me, "I was the only woman in my class, and one of only two women in the whole school. The other one was my sister."

"Did she go into pediatrics, too?" I asked.

"Oh no. Internal medicine. She was President John F. Kennedy's physician. But she died several years ago."

"Do you travel a lot?" I asked her.

"I try to. But it's been harder recently. Luckily, I learned years ago how to avoid jet lag."

"Really? How?"

"I stand up every hour, walk around, and have a nonalcoholic beverage. You'll be surprised, it works."

I talked to Betty for a few more minutes. Walter was now sitting on the armrest in front of me, talking to a middle-aged woman in jeans and tennis sneakers, with wrinkled tanned skin and curly salt-and-pepper hair.

"This is Bob," Walter said, reaching his arm out toward me. "Suzanne here is also from the Bay area."

"What brings *you* to New Guinea?" I asked her.

"I divorced several years ago. My kids all grew up and moved out, so I decided to embark on a new life. Now, each year I go somewhere exotic. Last year was Tibet, the year before that Patagonia. Three years ago I went to the Galapagos. This year is New Guinea. I scrimp and save all year to go. By the way, have you met Mark and Steve?" she asked. In the row in front of her, two men sat, bandanas around their heads, wearing unbuttoned plaid flannel shirts and tee shirts. I had noticed them checking in at the airport with long oars.

"We're going rafting," they explained.

"Rafting?"

"Sure. We've been to most of the world's other great places for white water rafting. Last year, the Amazon. New Guinea's the only place we haven't been."

"It sounds dangerous," I said.

"That's part of the fun."

"Have you been doing this a long time?" I asked.

"A couple of years. Mark just started medical school last year, so we've had less time. When we do have the time, we really go."

The three of us were the only people in our 20s who weren't missionaries.

"Where are you staying in Port Moresby, by the way?" I asked them. I didn't have a reservation or know of any hotels. In my excitement and naiveté, I hadn't thought about where I would stay—or even what I'd do in Moresby. I had planned a few days there, figuring that as the capital, it would be important to see. Carleton had been too busy to give me mundane details such as where or how long to stay there. I was on my own.

"We've arranged to stay at the medical school dorm. We wrote away in advance. Since the school's on Christmas break, they thought they'd have lots of space. It's also cheap. We're only staying one night, and are flying out tomorrow morning to start rafting. But I'm sure you'd be welcome to stay there, too. After all, you're here to do medical research."

By now Betty and her daughters were chatting with Suzanne. Soon we were all standing in the aisles conversing as if at a cocktail party. Our respective adventures somehow bonded us. "What a happy flight this is!" Walter exclaimed.

The only people still seated were a family in front. A middle-aged couple and their two daughters all sat reading books. Walter walked over and started talking with them, too.

"Have you met the Whittiers?" he soon called out to me. I strolled over. They were from Los Angeles.

"What are you going to be doing in New Guinea?" I asked.

"Every Christmas we travel. New Guinea is the most unusual place in the world we haven't been."

They asked me where I went to school. I told them. It turned out that friends of theirs—film producers who lived in Beverly Hills—had a son in my class. The Whittiers and I were glad to find a common connection. There, above the middle of the Pacific, thousands of miles from anywhere, we ended up talking for an hour or two.

A tall man with glasses and a long, greying beard stood to one side observing us all. I eventually walked up to him and introduced myself.

He was Richard Balsam, an anthropologist, was traveling to study tribal law. He asked about the research I would be doing.

"You must be very excited," he said.

"I am—but a little nervous, too. I've never done anything like this before."

"Have you studied a lot of anthropology?" he asked.

"Some." It turned out that we had both had the same professor— Clifford Geertz—who had taught Richard at the University of Chicago and me years later at Princeton.

"It's probably good you haven't had too much formal study in anthropology per se. You'll see phenomena more clearly. Your mind will be less muddled with theories. Just remember to keep in mind the most important lesson in all of anthropology."

"What's that?" I asked.

"The map is not the territory." I was a little puzzled. "Everyone makes up his or her own map of a place," he explained, "based on his or her perceptions and experiences. We all have different maps."

The only passenger in first class was a tall woman wearing dark sunglasses, an all-black silk pants outfit, gold bracelets, and several rings. She peeked back at us, seemed bored by herself up front, and soon came to coach class to join us. She was Australian, returning to visit her husband, who owned a company strip-mining the New Guinea rain forest.

"You know who comes to New Guinea, don't you?" she asked me, looking around us. "Missionaries, mercenaries, and misfits." I knew which group she belonged to, but wondered about myself, not liking the implication.

Eighteen hours later we landed in the capital, Port Moresby. As we all waited for our baggage, I ended up standing with Richard. A ground crewman finally brought out the luggage, piling it on the floor. There was no rack. A few baggage carts lay scattered about—all broken. None had intact baskets or were painted. I passed a counter labeled "quarantine" and, partially obstructed behind a desk, a sign proclaiming, "Malaria is endemic in New Guinea. If you get sick within six weeks of your arrival, contact a physician." My head pounded from the heat and the antimalarial medication I had already started.

We took our bags and proceeded through customs. Officials stopped only the woman from first class, wheeling a cart piled high with boxes of expensive gifts, and the two raftsmen, Mark and Steve, whose huge backpacks stuffed with sleeping bags and inflatable equipment towered over their heads. The remainder of us passed a partition and entered the rest of the room at this one–building airport. The walls were all painted dull yellow. Fans churned slowly overhead.

"My sponsor's meeting me here," Richard told me, searching around.

A short Chinese man walked up to us, accompanied by a tall blond woman in a lemon yellow Chanel dress. "Are one of you Richard Balsam?" he asked us.

"I am," Richard said, smiling.

"Larry Chu," the man said to both of us. "This is my wife, Diana," he added.

Richard, in turn, introduced me. We all shook hands.

"And what are you going to be doing here in PNG?" Larry asked me.

"Studying kuru in the Highlands."

"Through the Institute of Medical Research?" he asked.

"Yes," I answered, surprised he knew.

"Fine outfit," he said. "First rate. Do you have a place to stay here in Moresby?"

"I think so."

"If you have any trouble, please let me know. Do you have plans for dinner tonight?"

"I don't think so," I said, knowing no one in the city.

"We would be delighted if you would join us."

"Really? Are you sure that would be alright?"

"Absolutely. We would be honored." He gave me the number of a taxi company—the only one in the capital—along with his address.

Mark and Steve finally made it past customs, and the three of us took a cab to the dorm—a long line of rooms between a lawn and the jungle. We each got a room.

They took a nap, and I decided to tour the town.

No road left the hot, humid city for more than fifty miles before ending abruptly in the jungle. To get here from anywhere else in the country required flying. This nation, emerging from the Stone Age, depended more than any other on planes. Many natives had collected together enough money over several years to fly to the city, and came looking for work, only to find none. Stuck there now, unable to afford airfare home, they fell into poverty and resorted to thieving. The homes of the Chus and other expatriates—or "expats"—were all robbed at least once a year.

The town curved around a harbor. On heights jutting out into the sea, rambling mansions sprawled, affording expansive views of the water. Yet up the sides of the hills climbed shanty towns, and abandoned concrete World War II pillboxes, now used by prostitutes at night. I took a public bus inland to see the country's Parliament building, located at the end of a muddy dirt road. The national emblem had fallen off the side of the whitewashed stucco structure, leaving a grey silhouette. Nearby stood the National Museum, which, a sign said, was "Closed indefinitely due to leaky roof." I entered the Botanical Gardens across the street. A German couple joined me. But torrents of mosquitoes attacked us. We ran out as fast as we could. "We were in Australia and thought we'd stop over here in New Guinea on our way back to Germany," the man explained, "since we've never been here. But we just changed our airline tickets. We've been here for four days and it's been awful and depressing. We were going to stay here for two weeks, but now we're fly-ing back tomorrow morning. We'll be glad to leave."

I returned to the medical school dorm. I felt like a total stranger here, missing friends and family. Nothing was easy or orderly. I con-stantly had to make difficult decisions about and assessments of what to do.

I decided to walk around and see the hospital near the dorm. Thick mud oozed along the cement walks between the squat buildings. Inside the hospital, a woman lay asleep on the ground beside her ill child. Pa-tients all walked barefoot, not owning shoes. On the walls, old UNICEF posters faded, with paragraphs in English that the women could not read. On the doors hung primitive drawings of the male body, with en-larged, emphasized genitalia.

That night, I dined with the Chus. I met them at their house, a building on a cliff with a view of the water. An eight-foot-high black metal fence surrounded the property. Their 16-year-old daughter, Jen-nifer, joined us, wearing a fashionable yellow wraparound dress, a yel-low pin in her hair, shiny gold shoes, and red polish on her finger and toe nails. She had grown up here in Port Moresby and now attended a boarding school in Australia, from which she was home for Christmas vacation.

"What did you do today by the way?" Diana asked me.

"I went to the Parliament building."

"But how did you get there?" she asked, surprised.

"The bus."

"Really?" Jennifer asked. "I've never taken one. I wouldn't even know where they go. Don't mostly nationals take the bus?"

"Yes," I said.

"I deal with nationals as little as I can." The Chus called natives "nationals"—more politically acceptable, but equally, if not more, dis-missive.

"What did you do at the Parliament building?" Diana asked.

"I took some photos."

"Why?" Jennifer asked. "Isn't it falling down?"

"That's the whole point," I said.

They didn't understand.

"We're tolerant of any religion," Mr. Chu told me, changing the subject. "What religion are you, for instance?"

Shocked but wary of this social situation, I answered. "My mother is Jewish."

"I am not at all anti-Semitic," he said. "Let me tell you what I think of the Arab–Israeli conflict—not that I'm anti-Semitic. I think the two sides should be friends, and see their true enemy is the Bear—the USSR. But the Arabs are stupid."

"I don't know if I'd call them that."

"Why, I don't mind if even both your parents were Jewish," he explained.

Diana turned to me. "He's obviously not hung up about it!"

"My wife's best friend from law school was Jewish," he added.

"Maybe I know him," I said.

"Bullshitting like the natives do, I see," Diana said.

"I've never met a Jew before," Jennifer interrupted.

"We're really no different than anybody else."

"If you try to keep this hush hush," Mr. Chu continued, "I'll try to arrange a meeting for you with your minister—the Minister of Health."

"I don't know if he needs that," Diana said.

We left their house and drove to a Chinese restaurant where all the ingredients were imported.

The Deputy Minister of Justice, a "national," joined us. Larry spoke to him privately, then the minister turned to me.

"You Jew?" he asked, obviously tipped off.

"Yes," I answered, "Are you?" He wasn't sure how to respond.

Afterward, I returned with the Chus to their home to watch the latest videos from Hollywood—the major form of entertainment, I soon learned, among these expats and their friends here in Moresby. The capital seemed culturally dead, the most culturally deprived area I'd ever been in, lacking any signs of Western or native artistic culture—except for the painted hospital doors. Of the several expats I visited before eventually leaving the town, none had any Western or New Guinea art in their homes—not even reproductions. I missed even the cheap posters of Monet paintings back in the States.

"We are flying to Lae and then driving up the Highlands Highway," Larry said. "Why don't you join us? We can drop you off in Goroka."

I wanted to be on my own, but to see more of the country, too. By joining them, I would get to visit Lae—where Amelia Earhardt was last seen—her last takeoff point from Earth. I decided to go.

The next morning, I woke up in the dorm and went to the building that housed the bathroom. The open air structure had no doors or windows—just open portals for the air to flow in. As I stood shaving in front of the mirror over the sink, I suddenly noticed something move over my head. A lizard—like a baby alligator—was crawling up the wall just behind me. It was brownish-black and eight inches long. I froze. It did, too. Slowly, I turned around. It stayed perched on the wall, darting its tongue in and out. A gecko, I decided, probably harmless. Still, I finished shaving as quickly as possible, and hurried out.

Later that morning, we left the capital and flew to Lae. The town, once the third largest airport in the Pacific during World War II, now, fifty years later, still consisted of only a few metal buildings and wooden shacks built around the airstrip. The jungle encroached.

That night for dinner, the Chus took me with them to visit Sam Chan—a relative of the Prime Minister, Sir Julius Chan. Sam's father had come from China with his two brothers. Prohibited from bringing women with them, they had married natives. Sam now owned a store next to his home. All but two of the trade stores here were owned by Chinese. The other two—the largest, Burns Philip, and the Steamships Company—were owned by the Australian shipping lines that transported the goods, allowing them to dominate the market more firmly.

"I want more Americans to come to New Guinea," Mr. Chu said. "Australians control too much." His eyes narrowed. "We need other Whites besides Australians here."

Other ministers of the province came by. "Locals, not expats, should be in charge," they all agreed. "But the old form of authority, village law, is breaking down." The implication: expats still needed to have authority.

PNG had become independent from Australia in 1976. Since then, two political parties—that of Michael Somare (a national, and the first prime minister) and that of Julius Chan—have alternated power. Politics has revolved around the rise and fall of these two groups, each coming or going.

"The national government, now under the opposition party, does not have the interests of the people in mind," Larry now leaned forward to explain to me.

That night I went to bed in the town's one motel. I felt alienated, knowing no one in this bizarre country and disturbed by some of the attitudes I was encountering. Why was I here? What was I realistically hoping to do? Was I wasting my time? I reminded myself that I had traveled here to observe and learn, and would eventually be moving on. I thought of all that explorers, scientists, and anthropologists must have put up with. 'Just relax,' I told myself. 'This is an experience I'll probably only have once in my life.'

In the morning we started up the Highlands Highway—the only road called a highway in the country—which went from the coast into the mountains. Despite its name, the Highlands Highway was merely a thin dirt track ploughed through the jungle only several years before my arrival—a mere thread of dust when rain wasn't falling. The road was the only one leading from the coast into the still partly uncharted Highlands. No Westerners had ever been to some parts of this mountainous country. Official government maps still depicted large areas merely as blank, pale green stretches labeled "uncharted." I looked back at the few metal shacks of Lae, beside the Pacific, feeling I was leaving something behind.

The road was primarily one lane. When another vehicle came, we had to pull over to the side. Yet this was the lifeline for most of the country and its population.

At the first mountain pass, a landslide had caused a traffic jam. Along the twists of road, huge freight trucks and passenger-stuffed pickups waited. The natives all sat nonchalantly, as if they had all the time in the world. Our small car by itself squeezed past the larger vehicles and continued on.

I looked out the window at the rain forest around me—the first I had ever seen—far denser and lusher than I had imagined, the trees and leaves all unusually shaped. We rumbled further into the mountains, their pale blue peaks cloaked by dense, swirling clouds. This road fragilely linked two worlds that could not have been more different. I little knew where I was headed.

The day was long, driving on this dirt track. Finally, we stopped for a late lunch at the one restaurant we passed, and had beer, hamburgers, and potato chips—all that was on the menu. Eventually, we arrived in Goroka, which was to be my only tie to the outside world during the upcoming months. Here was the last outpost of the world—as far as telephone, electricity, mail delivery, and imported goods stretched. None of these things existed north or south for hundreds of miles. The town lay in a narrow valley, rimmed by tall mountains, beyond which even steeper ridges rose—a series of pale turquoise veils. Against this vast backdrop of ranges, a few strange trees poked up, sprouting leaves only at the very top. A carpet of luminous light-green grass rolled between the edges of the valley, stopping abruptly at the foot of the hills.

Compared to Port Moresby, the West had penetrated and influenced the Highlands much less. Expats were fewer. The air seemed cleaner and clearer, and poverty absent. After all, for millennia the Highlands had been self-sufficient, supporting its whole population.

The town's very existence seemed strange, almost miraculous. Barefoot natives wandered along the main street by the few wooden stores—the one bar, one small hotel (called the Bird of Paradise), one bank, one general store, and one bakery (the only place for hundreds of miles around that commercially baked bread and cake). An isolated loneliness pervaded the air—a sense that this was, in fact, the last outpost of civilization. The town had the feeling of the frontier, of the American Old West. Here, in the midst of a vast uncharted wilderness, the few stores stood as tiny reminders of the outside world.

Initially, the town had just been an airstrip—a single dirt runway surrounded by tall grass—the only connection with the outside world. Around it, these few structures had slowly been built. The buildings remained centered around the airport, just as cities in the past had been built along the sides of rivers. Only later had the highway been built.

The Chus dropped me off at the Institute of Medical Research—a compound of dark wooden buildings on top of a hill. I carried my bags up the driveway, opened the screen door, and walked in.

I knew almost nothing about my new boss, the Director, Dr. Michael Alpers—my only contact in the country—and was apprehensive about meeting him, given my youth and relative lack of experience. I hoped I would get along with him. My exposure to expats thus far had left me wary.

The only person in the building was Michael's secretary, Ami. "Dr. Alpers is away for two weeks," he told me, "Dr. Alpers told me to give you the keys to his house and to take you there. On Monday we will drive you out to Waisa where you will stay." It was now Friday.

"Did Dr. Alpers leave any instructions for me?" I asked.

"No. You will go to Waisa to stay with Roger and Maryanne Richardson who are building washhouses for the people. Your instructions will be waiting for you there. I assume you will be helping them."

"Building washhouses?"

"Yes. They will have further instructions for you." I felt disappointed. I had come all this way to build washhouses?

Ami gave me the keys to Alpers' house and dropped me off, quickly driving away.

The ground floor of the house contained only a garage. Metal stairs led up the side to a veranda and presumably the front door. I started to climb up. All around me, spiders had spun webs, cloaking the staircase in a dense tent of woven silk. At a few points I had to duck to avoid scraping my head on this canopy. In the middle of the silken nets sprawled fat spiders—black and fuzzy with blue spots, shiny and yellow with red dots—many the size of plum tomatoes and looking like tarantulas. The variety stunned me; I hoped none were poisonous. The webs stretched under the eaves of the roof as well.

I opened the door and walked in. From the walls hung long pointy spears and wooden shields carved with alligators and human forms. The carvers had made the shields crookedly, not along right angles (and didn't show much of an attempt to conform). In the centers floated vaguely humanoid shapes with protruding spirals, suggesting arms, legs, penises, and testicles. All of these slanted at odd degrees. None stayed rigidly horizontal or vertical. (Nothing in the human body in fact does, I realized.) A glass-topped box held a collection of butterflies, with wings of bright electric blue, and gold tinged with ochre. In another box, seashells spiraled in all shapes and sizes. White spots, resembling tiny leaves, decorated dark greenish-black shells. Two large old-fashioned, hand and pedal-operated looms—presumably belonging to the director's wife—filled the center of the living room. Purple, yellow, orange, and red wool blankets, half-woven, were rolled down, the colored yarns dangling off each side. Floor to ceiling bookshelves held literary classics from around the world, including books by Confucius, Cervantes, Thoreau, and Proust, along with a collection of old classical records—the complete works of Bach and Mozart, though almost nothing else. I had brought along only two books to read and there were no bookstores for hundreds of miles around. Alpers' house reminded me of all I had left behind in the States. Part natural history museum and part Shangri-La, his home was a welcome oasis.

I would be staying here for the weekend by myself. In the morning, I wandered out to see the town. The few streets were all unpaved. The tallest building was two stories high. Most were single story. A man walked down the street, chewing a three-foot-long stalk of sugar cane. The pale wet white fibers stuck out of his mouth as he sucked out the juice. He spat the pulp on the ground. At the market, women crouched on the dirt, each behind colored sheets, displaying a few bunches of bananas, coconuts—the shells encased in even thicker outer husks—or betel nuts. The bananas were of several kinds, some short and stumpy, others longer, resembling American bananas, and more expensive. I bought the latter and excitedly peeled one. It looked like a normal banana, and I put it in my mouth. But I could barely bite it. It was like eating styrofoam. I threw the fruit into a garbage can on the street, and sat down on a bench. Some natives standing around watched and laughed at me. "Why do you throw out good food?" they asked me.

"They are unripe," I said. The natives laughed again.

"You tried to eat these?" They chuckled. "They are good. But only for cooking." Only the short bananas were edible raw.

I walked around further. I bought bright green betel nuts, which I tried eating. They had been dipped in lime powder. I had heard that natives chewed the pulp, then spit it out, to give them a high. But the fruit sucked all the moisture out of my mouth. I felt a little spacey, but didn't know if I wasn't just tired, hungry, and overwhelmed. It was a different, though not particularly pleasant, sensation.

Back at Mike Alpers', I noticed a phone on a corner table, and tried calling home collect.

Static crackled on the line. Finally, a New Guinean voice spoke. "Hello? Hello?" On my third try, I got through. My mother answered. I was relieved. "I got here okay," I told her. "But it's hard being here. I feel I don't know anyone."

"Of course you don't. You just got there," she reminded me. "Just remember: if you don't like it you can always come home." It was reassuring to hear that, but I also knew that I wasn't about to leave. Despite the loneliness and difficulty, I wasn't going to quit now after coming this far.

I was sitting in a comfortable modern easy chair by the phone when I heard a tinkling noise. It got louder. I got up. The wine glasses in a sideboard beside the dining table were jingling; the plates started to rattle. Oh my god—I suddenly realized—it was an earthquake. I had never experienced one before. What was I supposed to do? Lie on the floor?

Sit in a chair? Get out of the house? For several seconds I stood dumb-founded. Then, the vibrations started to die down. It had only lasted a few seconds. My breathing and heart rate began to return to normal, too. The shaking stopped. Here in New Guinea, part of the Pacific Rim, earthquakes frequently occur.

Two days later, Ami supplied a driver, named Kanaua, with whom I set out for Waisa on a side road, even less developed than the High-lands Highway. Heavy rain poured down, galloping on the canvas roof. The windshield wipers sloshed back and forth, painting beige strokes across the dirty glass. I bent down and squinted to peer through a few curved streaks of thinner, more watery mud, but couldn't see where we were going. Outside, rain and mist obscured the landscape. Water seeped in along the tops of the windows and streamed down beside me, wetting the seat and my clothes. I pulled into the cab as far as I could, as a shiver pierced down my spine.

The road slithered over the fingers of mountain ranges rising up all around. Along the road, steep mountain walls tumbled down. Clear wa-ters splashed down ochre, grey, and red clay slopes, between smooth rocks. From the colored clay, stalks of tall, lush green grass sprouted. Thatched huts huddled under surging trees. Jagged mountains lifted up and over each other, ever more pale and distant, their ragged tops like ripped sheets of blue paper against the sky. These hills had risen through the effects of continental drift as Australia, pushing north, rammed into New Guinea and crumpled up the earth like a crunched-up piece of pa-per, creating the mountains that now ringed us on three sides.

This thin road wandered into the remotest stretches of the inhab-ited earth. Few, if any outsiders had ever traveled to some of these areas. Some children had never seen a White person. Landslides peri-odically washed out several sections. We passed Kuru Mountain, rising tall and wide, separating the Fore and neighboring groups from the Goroka Valley and other regions, and creating seclusion for millennia. Kuru had flourished on the far side of the mountain and been absent on the other. No other area in the world has been named after a disease. Plague, syphilis, tuberculosis, and cancer have lent their names to no place—except hospital wards. Such was kuru's terrible impact.

The mountain stood now, its peak rising against the darkening sky—a wall that would lock me into my new home for the next several months. Until the late 1950s when this road was built, the Fore had no word for what lay over this mountain—because no one had ever been there. Traditionally, to get to Goroka Valley from the nearest Fore village took several days by foot, with the trekker having to carry more than one-third of his or her weight in yams as food.

We passed a few small groups of people along the road, who sang out "Ai Ai Ai Ai Ai," and waved—differently than in the West, with fingers spread out vertically, a sideways "royal wave." Children shouted in Pidgin, "Hi, masta," from the English, "master," a term used to address White men—its connotations apparently unknown.

After several long hours on the road, Kanaua pulled over and stopped the car. It was now pitch dark. A heavy rain pounded on the roof. I looked out and saw nothing—only blackness and woods. Not a single light.

"Where are we?" I asked in Pidgin English, which Kanaua used to communicate at the Institute, and the rudiments of which I had learned from Carleton's adopted children, before I left the United States.

"Waisa."

"This is Waisa?" I asked in shock.

"Yes." My heart sunk.

He got out. I took a deep breath and then followed his lead, running to the back of the truck in the torrent to grab my two bags from under a tarp. They were soaked. He ran ahead and I followed, splashing down a path until we came to a small hut with a porch. I stomped my feet on the wooden porch to dry them off, and he opened the door.

"This is Bob," he said, introducing me inside to a tall man with a light brown beard and to a short woman with long, dark brown hair.

"Roger Richardson," the tall man said. "This is my wife Maryanne." She glanced up from the ground and nodded reservedly. "You got here okay?" he asked. His voice, with its Australian accent, surprised me in the middle of this wilderness and it was unusual to my ears. Yet it was refreshing to hear English.

"Yes," I said. My wet clothes clung to my body. "I made it."

"Me go," Kanaua suddenly said. I turned around. Just as quickly as he had come, Kanaua fled. His running steps, stamping in the mud, disappeared into the splattering rain. He had a long drive ahead of him to Goroka. I was now here alone.

"Let me show you around," Roger said. "This is your room." He directed me past a thin wooden door to a narrow space containing a cot, a desk, a book shelf, and a small window. Some walls were of rough-grained wood, others of plaited bamboo. I would be living inside a woven basket.

"That's our room." He pointed to another door. We didn't enter. "This is the kitchen," he said, directing me to a small area with a sink and a cupboard covered with metal mesh, "to keep the rats out." "This is the wash area," he said as we passed a small counter and sink in the back. A tiny shaving mirror tilted against the wall. "Finally, this is the shower room." In a small closet-like space, a hook hung from the ceiling. But there were no taps or running water. I looked confused. "You fill a bucket with water," he explained, "hoist it up on the hook, and tilt it to pour water over yourself."

"I see." I cleared my throat.

"For the bathroom," he added, "there's an outhouse down the path." We were standing in a central area beside a table and a black, cast-iron wood-burning stove. That was the whole house. It lacked electricity. A hissing Coleman kerosene lamp provided the only light. The walls were empty. I tried to be civil, but hadn't expected as much barrenness. I felt I was in the nineteenth century Wild West. Later, I learned that the house also contained a flashlight and a short-wave radio receiver.

I would soon find out that it was just about the only house for countless miles around that was rectangular (as opposed to round), and had a metal (as opposed to thatched) roof. The roof was an angled sheet of corrugated grey metal. A gutter along the side collected and emptied rain water into a drum, from which spouts poked into the sinks, supplying water. Traditionally, the natives drank from streams in which pigs and wild boars shat. Intestinal parasites probably infested everyone who drank the water.

As a health measure, the Institute had arranged for Roger and Maryanne to build washhouses—small, yard-square structures with smaller corrugated metal roofs, collecting water into smaller drums. Michael also had the Institute build our hut as a research base.

We now sat down at the table in the center—I on the right, Maryanne in the middle, and Roger on the left.

"Do you have instructions for me?" I asked Roger.

"No," he answered, surprised. "I thought you would be working on kuru here." I was relieved that I wouldn't be building washhouses. But I still lacked specific instructions. "In fact," he continued, "Carleton's two guides—Sana and Sayuma—will be by in the morning, raring to go. But be careful. It's very hard getting anything done here in the jungle. When we arrived two years ago, Mike Alpers had only one piece of advice for us: expect only failure. He was right."

I hoped he'd be wrong. But when I got into my room and closed the door behind me to go to sleep, Michael's words echoed in my head.

PART II BUSH ROUNDS

A Man of Crossing Rivers
on Trees

In the morning, Maryanne made coffee on a small kerosene burner. We were seated when we heard a knock at the door. Roger went to open it.

"It's your guide, Sana," Roger said. I stood up and walked outside. Sana was short—the top of his head reached the middle of my chest. Yet his chin rose in the air, and his cheek bones were high; his eyes glittered and he had fine features. As a walking stick, he used a black umbrella topped by a carved, shellacked wooden handle—an odd accessory here, right out of the City of London. It was as much an artifact here as a New Guinea mask would be there. Yet it lent him a distinguished, almost Edwardian air.

"Me come," he said in Pidgin.

I wasn't clear what to say in return, not sure how to say "nice to meet you" in Pidgin, or even if one could—if such words existed in the language. But he seemed pleased to meet me. "You pren bilong Carleton," he added (Pidgin English for, "You're Carleton's friend").

"Yes," I said. That was important for him. In this culture, lacking bureaucratic or other formal institutions, the fact that I knew someone whom he knew gave me entrée. In his mind, I would later learn, I was Carleton's *wontok* (from the English "one talk" and meaning part of a kinship line).

That afternoon, as Maryanne and I were sitting in the house drinking tea and having strawberry jam on bread, we heard another knock at the door. Maryanne answered it. "It's Sayuma," she said. A thin man pushed his way past her into the house. He had an angular bony face, and a scar about his right eye, and was missing most of his teeth. A few remained, yellowed and exposed down to their roots. His gums had rotted away from extensive use of betel nut, I later learned. "I am your guide," he announced.

"Yes, and Sana, too," I said.

"Yes," he said reluctantly, "and Sana, too."

We made a plan for Sayuma to take me the following morning to see my first kuru patient—a few hamlets away. Sayuma himself was from Karamuni hamlet.

"Should we contact Sana?" I asked.

"I will let him know."

"Really? . . . How?"

"Someone from his line is now here in Waisa and is returning today to Purosa, where Sana is from. This man can take the message for Sana to meet us in the morning. What time would you like to go?"

"As early as possible, I suppose," I said, surprised he even discussed and knew about time. "What's good for you?"

"You name a time."

"Okay—11."

"I think you should go earlier," Maryanne interjected. "The rains start everyday by midafternoon. Roger leaves the house every morning at 7." Indeed, I would soon see that during the rainy season, lasting six months a year, showers usually fell every afternoon and evening—though the equatorial sun scorched the earth by mid-morning.

"Okay, how about 8?" I said, thinking of Sana having to trek here all the way from Purosa, which I figured would take a few hours.

"Okay," Sayuma said, very confidently.

"You're sure that's alright for you?" I asked.

"Yes." He nodded. He got up to go, taking a piece of bread with him.

But by 9:30 the next morning, my guides hadn't yet arrived. Finally, Sayuma showed up at about 9:45, Sana close to 10. I brought a knapsack containing a camera, a notebook, a pen, and a bottle of water for us. My guides brought only handmade bows and arrows, which they carried now. I didn't know why.

They took me to see the patient. On the road we passed lush green valleys, and blue mountains that lapped over each other like waves on a sea. Huge bamboo trees rose up elegantly, the tops shaggy and dense

with thin green blades. Trumpet flowers hung down vertically like car-
illon bells. Thickets of sugar cane shot up seven to eight feet tall like
giant sheaves of grass, sprouting plumes on top. Compact bunches of
bananas hung from trees. Smooth, curved lime-green fruit, each perfectly the same, seemed almost unreal. The leaves, as if grown too long,
tumbled over fences. On the sides of the road, delicate ferns rose from
crimson clay. Flowers with black centers and yellow-orange petals (cos-
mos, I was later told, escaped from the gardens of missionaries in the
towns of Goroka and Kainantu) sprouted wildly. We passed women carrying huge loads on their heads, balanced with their hands—bushels,
bags, whole tree trunks.

After two or three hours, we turned into the woods, along a narrow, barely discernible trail. Deep rain forest soon engulfed us, miles
from any other human being. My God, I suddenly thought, I'm here
with a bunch of cannibals, wholly at their mercy. My New York City
'street sense' made me cautious.

Cannibalism is easy to demonize. In the West, it still represents an
ultimate taboo, a violation of human dignity. Rumors persist in the
United States that Michael Rockefeller had been eaten—which is not
true. The practice has a long history of lure and lore. Even Montaigne
wrote an essay on it, urging acceptance. It appeared in the press most recently in association with Jeffrey Dahmer, who added sexual abuse. Yet
it happened here in New Guinea and needed to be accepted, not just
morally condemned. My guides seemed trustworthy, not aggressive. In
any case, here in the middle of the forest, I had no choice but to trudge on.

Suddenly Sana turned to me, laying his finger over his lips, and
motioned for me to stop.

"Wanem i pass?" I asked—Pidgin English for, "What's going on?"

Sana shook his head at me and turned away, not speaking. Did he
not understand or not want to answer?

"Wanem?" I repeated. Still, Sana wouldn't respond. I stamped my
foot.

"Bird," he finally whispered.

"Bird?"

Sana pointed. I looked in the direction he was indicating, but could
see nothing but towering dark trees and thick vines. Little sky peeked
through. I certainly didn't see a bird.

Slowly he slipped out from a bamboo quiver an arrow that he positioned across his bow. He pulled back on the string, aimed, and silently released.

Squuaaaaakkkk! A beautiful red, yellow, and white-plumed bird, pierced by the arrow, tumbled suddenly from a branch high above and landed two feet in front of us with a thump.

I have mild red-green color-blindness, and slight nearsightedness—though I can drive a car without glasses. This problem had never interfered with my life or functioning in the United States. But, I suddenly realized, I would not have survived here in this hunter-gatherer tribe. Color-blindness is much more common in industrialized than in primitive cultures, where, I now saw, it confers a selective disadvantage, and was bred out. As a result, my guides possessed survival skills that I lacked. Human culture thus shapes genetics and evolution.

We continued climbing the muddy trail. My feet kept sliding, and I invariably lagged behind. At home, I jogged three times a week, but was out of shape compared to my companions, who commonly walked twenty miles a day, up and down mountains.

As we hiked, Sana pointed out: "We use this tree for building huts, this plant as food, these leaves as dyes. This plant," he said pointing to a vine, "we do not eat. It is a no-good plant."

On arriving in the hamlet, I felt awkward, but played it cool. My guides and I shook everyone's hand. Sana and Sayuma sat down quietly and motioned for me to do the same. Nothing else was said or done. Smoke rose from the thatched roofs of huts—fires apparently burned inside. Blue smoke seeped through the grass tops. I was surprised the huts didn't catch fire.

The majority of the people in the region lived in round, thatched huts grouped into hamlets—though some stayed in the bush, and were said to be "not men of clearings." This hamlet—like most, I would soon see—consisted of three or four huts. Groups of hamlets, usually not more than an hour's walk from one another, constituted villages, which in turn could be an hourslong or even a full day's walk from one to the next. Several dozen "villages" usually formed their own cultural and linguistic group.

After several more minutes, Sana started to talk to the men in *tok ples* (Pidgin English for "local language"—from the English, "talk place"). They all still looked down at the ground between their feet. After a long conversation, Sana glanced up and nodded at me. It seemed my cue. I explained in Pidgin English that I was a doctor from America, here to examine kuru patients. My guides, called *turnim toks* (Pidgin English for "turn talks" or "translators"), translated Pidgin into the local language. Pidgin was the national language, given the number of disparate tongues spoken on the island—a result of the steep mountains isolating each valley, leading to the development of distinct cultures. This *lingua franca* combined elements of French (e.g., *save*, pronounced "sav-ay," from the French "savoir," meant "to know"), German, Dutch, English, and Malay. Parliament was conducted and newspapers were written in Pidgin. I had bought a dictionary, *Buk Bilong Tok Pisin* (The Book of Pidgin English), but communication remained difficult.

I explained now that I wanted to see the man's wife. Sana helped me out, explaining where I was vague. When I finished, there was silence. I didn't know if they would agree. Slowly, the oldest man in the group began to talk. Sana listened, turned to me, and nodded.

"Can I go in the hut now?" I asked him. He nodded again. Slowly, he stood up. I followed his lead. I had never examined a patient before and felt nervous. But I had seen Carleton's many films of kuru cases. I knew from these movies what kuru looked like at each of its several stages.

I walked solemnly behind Sana into the round hut, and ducked through a small door into a cramped, dark, smoky space. White ashes smoldered on the ground in the middle. Light barely seeped through the cracks in the woven bamboo walls. Smoke stung my eyes. The hushed blackness swallowed me. I could barely make out the body lying on a platform. Sana told me to sit alongside the patient. He then squatted down beside me.

I was afraid to touch her. I knew that infected brain matter contained the highest concentration of the virus-like agent, but what if her hands were contaminated with her saliva, urine, or feces? Could I get the disease? I shouldn't have been afraid (it wasn't scientific). But I was—even to admit to myself that I was afraid. Discoveries about infectious diseases often require taking risks.

Yet Carleton and Michael had examined these patients. Perhaps these researchers had become immune over time, though I doubted it.

Slowly, I lifted her hand. It rattled back and forth in the air. I asked her to hold her hands up, but she didn't respond. I called her name. She didn't answer, and kept staring straight out in front of her.

I wanted to do other tests of coordination, and asked Sana if I could take her outside. He shook his head. She was too close to death.

I counted the fast breathing of her still body, and her pulse. This woman, Wanebi, had engaged in cannibalism, and was now dying of kuru as a result. "Is there anything else that I could do—that Carleton or Michael would do?" I asked Sana.

He shook his head.

I realized how little I could offer—certainly no treatment—and how limited scientific understanding of her disease remained. I felt terrible for her. Seeing her lying there, I had a crushing, suffocating feeling in my chest. For a moment, her illness gripped us both, albeit in very different ways. Death had always made me feel cold and eerie. We are all destined to die, yet it is still tragic when someone does. I had known well only one person who had passed away—my aunt, my father's sister, who had a massive heart attack and had succumbed the year before. The suddenness had stunned me. Instantly, she—who had hugged me tight every holiday I could remember since growing up—was no more. I sensed a gulf. One of my earliest memories was of President Kennedy's funeral—watching on black and white TV over and over again for days a riderless horse leading the dark, draped coffin through a cold, grey November day. Even recently, walking through a cemetery that overlooked an ocean gave me the creeps. It felt oppressive, claustrophobic, as if something were clutching at me. I wanted to escape the endless rows of faceless graves as soon as possible. I would die some day, too, of course. But though we know this, to permit ourselves to feel it takes hard work. Science offered some consolation—that through patients' suffering might come insights that could potentially help others through an increased understanding of the causes and treatments of these and other diseases. I could try to make something of the experience. But I felt uncomfortable just standing there now, watching Wanebi. I stepped out of the cramped hut into the dazzling equatorial sun, disoriented and confused.

I had brought her a blanket and a bar of soap, which were rare here. Her family took these now, and thanked me. It was a small token of my concern and appreciation. Her husband and her brother were seated to the side. I joined them. "When did she first become sick?" I asked.

"Last Christmas. At first we took her to a medicine man who gave her herbs. But it didn't work." We sat there for a few moments. It occurred to me that these older residents probably possessed other information that would be important.

"Did anyone else in her family have kuru?"

"Her mother, aunt, and oldest sister." I was surprised. But I wasn't sure what else to ask. My guides got up to go and I followed.

On the way back, I thought of the questions I could have posed—had she been at any other funerals and if so, which? Who else had been there? How many later developed kuru and when?

We returned to Waisa. The hamlet, I saw was surrounded by a rough, primitive fence of horizontal logs woven, using native vine rope, onto spliced vertical trunks and live plants known as *pitpit* (related to sugar cane and consisting of several varieties, of which the one used here was inedible). I stepped on horizontal wooden slats that rose up and down over the fence, serving as stairs into the compound. This primitive gate, an entrance, made the area feel like a home, surrounded and protected—though probably built to keep in chickens and pigs. Inside the house, Roger was talking with Soba, who rented to the Institute the land for the house. Soba had once worked as a doctor's assistant in Goroka. "This is Bob," Roger said to him now, introducing us. Soba turned his eyes slightly and nodded. He wore a red and navy-blue ski cap—covered with bits of vegetative matter, dirt, and flecks of loose white fabric. I would soon see that he always wore the hat. It was a trademark. He soon finished his conversation with Roger and left.

Sayuma asked me what time I wanted to leave the next day. Both Sana and Sayuma, I now noticed, wore plastic digital watches—the only ones for miles around. My guides sported these time pieces with pride. I would later learn that Stan Prusiner had given the watches as gifts. "Do you know how they work?" I asked.

"Yes," my guides assured me.

"It would be great to leave as early as possible," I said. We had a long trip ahead of us to another village to see a patient, and might not be able to do it all. "But I know this might be hard."

"No," Sana said. "You say what time you want."

We agreed on 7:00 a.m. But they arrived at 8:30.

I tried to teach them how to tell time with their digital watches. "Write down the time that we say we will meet." But they drew each number on the page as a separate figure, not connected to others, not part of a set of numbers. They made notations, but didn't see them as linked. They didn't think in terms of sentences—not seeing any written examples in their environment. They didn't read newspapers or books. Historically, writing started in trade—to keep track of goods. Here, too, numbers and names were all that were needed for the job at hand. Written verbs and sentences, I realized, must have been developed only later. My guides had worked with Carleton and Michael and wanted to appear to work like Westerners. But it was still difficult.

We hiked through the forest. To my surprise, I enjoyed being removed from civilization as we trekked on winding trails through the bush.

That day I visited a patient named Wasoru. Her family said she had had, and overcame kuru, but then could no longer walk without a stick. They, too, had tried native herbs, but to no avail. She clearly had kuru now and could stand only with handmade crutches, and still hobbled uncontrollably. I felt frustrated not being able to help her more, either. The Fore could try counter-sorcery treatments, but I knew they were ineffective when tested in the lab. It was easy to remove myself emotionally and just look at this situation clinically—it was, I would later see, what many doctors in the West did. I have to admit there was comfort in that approach. But it wouldn't help this patient or others I would soon examine. There was also some relief in knowing that I could leave, that I was only here briefly, observing, as a researcher. Yet that in no way lessened the awfulness of the problem she and her family faced. Physicians' distance allows them to go from one patient to the next, helping each. Yet such remove must be balanced with compassion.

I sat down with her family and collected additional information. Her mother and two sisters had also died of the disease.

"Who's feast was Wasoru at?" I asked.

"Many, many," an old man said. It was "time bilong kutim na kaikai" ("time of cutting up and eating").

"Over many years?"

"Yes." She had probably been exposed many times in her life. Everyone else in her line had also been to more than one feast.

Before we left the hamlet, an older man teetered up to me and offered me part of a long, oblong, orangish-brown gourd—a *karuka* or pandanus—hacked lengthwise into two pieces, and filled with seeds like a pomegranate. The whole fruit was covered in white and grey ash—evidently cooked by sticking it in a smoldering fire. I looked down at this odd, unfamiliar sphere in his hands. He sensed my uncertainty. He plucked out some of the seeds and popped them in his mouth to demonstrate, cracking them with his teeth and spitting out the shells. Then he thrust the dirty, ashen fruit toward me. Everyone looked up at me expectantly. I felt obliged to try, and scooped up some of the seeds and put them into my mouth. They were hard, bitter and dry. I bit the seeds to open them. The gritty, acidy skins irritated the walls of my mouth. Everyone waited for my reaction, smiling, their eyes wide, expecting me to be pleased.

"Em i gutpela?" Anyana asked, Pidgin for, "Is it good?"

I nodded my head.

They all smiled and relaxed and the man handed me the fruit to take with me.

I took it and thanked him.

As we hiked back, I asked Sana and Sayuma if they wanted some. They excitedly took it. I sensed it was special. I later learned it was a sort of delicacy here. I encouraged them to finish it.

On the way back, we took a short-cut and had to cross a river. In the rains of the past few days, the waters had risen, and were now too high and swift to ford. We hiked downstream until we came to a felled tree that stretched across from one bank to the other, high over the water. My guides were worried and didn't think I'd make it across. They wanted to take a several-hour detour to a bridge several miles further downstream. But it was already getting dark. "I will try crossing the tree here," I announced.

"No," Sana said. "We know how to cross here. We have done it before. But you are not a man of crossing rivers on trees. I think you want to go over the regular bridge." Sayuma shook his head, also thinking me foolish. But I was worried about the gathering darkness. The last thing I wanted to do was detour. I would try the log.

Sana first tiptoed nimbly across. As a kid, from the high bleachers of circuses at Madison Square Garden. I had watched my share of tightrope walkers. I gave my knapsack and camera to Sayuma, removed my footwear, balled up one sock in each boot, and held one boot out on each side of me, as far as my arms would reach. Then I tiptoed across, one step at a time, not transferring my weight from one leg to the next until each new footing was secure, balancing myself at all times with the boot on either side. Below, the rapids gushed. White furrows of water pounded the rocks. But I forced myself to focus all of my attention on each step. One slip and I could end up dead. The intense concentration felt very Zen-like.

At last I made it to the other side.

(Months later, when I developed my film, I found that Sayuma had taken a picture of me as I crossed. The photo was blurred. He had moved as I had, not knowing to hold the camera still.) Sana couldn't believe it and was overjoyed. "How did you do it?" he asked.

I wanted to explain that it was Zen-like. But Pidgin lacked the words. "It was the only thing I thought of," I said. My guides didn't understand. "It was nambawan," I added (from the English, "number one," meaning "the best"—the only superlative in Pidgin). They nodded politely. I couldn't begin to communicate how powerful it felt. I felt frustrated not being able to express what I wanted. Unfortunately, for the next few days I would talk mostly in this limited language. I felt mentally confined. The word units were small. Hence, any complicated terms were long. The news on the short-wave radio here took five minutes in English and twenty minutes in Pidgin. Yet the terms often astonished me. The term for piano was, "Big mama wantime plenti tit, yu hittim na em i cri tumas," from the English, "Big mama 'onetime' [meaning simultaneous or with] many teeth, you hit him and he cries too much." The term for bra was "Big shirt em holdim two titti bilong white mama." "Lukim you behin" meant "See you later." In newspapers, for example, Prince Charles was referred to as *nambawan pikinini bilong misis kwin* (the eldest child of the Queen). Pidgin itself is derived from "Business" English. New Guineans compensated by speaking very rapidly. The natives didn't think the words odd, not knowing English and hence the roots being employed. Yet if I left out even one syllable, the natives would either correct me or fail to understand me. To say "Kill" is not "Kilim," but "Kilim i dai," that is, "Kill him, he dies."

When my guides and I returned home after the long day of trekking, I invited them in. I served food, knowing they would eat whatever

I put out, but not wanting to be cheap. They each told me of several rumored cases of kuru, people we then made arrangements to visit on several day-long excursions from Waisa. We also planned a few longer, overnight trips to the edges of the region. "We will need other men and boys on the patrols," Sayuma said.

"Let's just go with ourselves for now," I said.

"No, take others."

He seemed insistent, but I remembered that Carleton's only tip to me was to limit the number of guides I took. I remained firm.

That evening, I walked to the edge of the ridge overlooking the fires of hamlets and the lush valley—a broad valley of bamboo and odd vertical trees that terminated in dense bushes of green high above the forest floor, seeking sun above all competition. Evolutionary biologists have shown that most species, from human beings on down, evolved here in the tropics and then spread elsewhere. Rain forests were places of possibility. I was inspired as to the wonder of being here—where men and women lived close to how they did when they started. Here was man himself.

Until only a few years ago the Fore had no metal, let alone cars or electric appliances. This group had been here for centuries, yet no buildings or structures indicated that people had resided here for more than a couple of years. No traditional house or structure lasted in the jungle, given the fierce downpours every day during the rainy season that extended almost half the year. The only evidence of the past extending more than a few years back were hillsides, mostly near Goroka, now covered with grass rather than trees, suggesting that slash and burn agriculture had once been practiced there. Some forests were also regenerating as a result. But when and for how long such agriculture had been practiced—whether 50, 100, 1,000 or 5,000 years ago—wasn't clear. (Only scientists using modern technology could perhaps piece such a history together.)

In the West, hundreds of years of scientific and technological development have expanded knowledge of the physical world and the range of possible possessions, activities, and desires. Yet here, there were few clothes and only a handful of crops. The Fore had no written history or pictorial art—nothing to distinguish anyone beyond their lifetime. Nothing except each other. I would soon see that they decorated their faces with colored dyes, squiggling lines across foreheads, and down cheeks and chins. Their only expressive artistic media were their own bodies. With little besides each other and their bodies, physical signs of affection were strong. At *mumus* (large feasts) and funerals, I would soon see, everyone gathered for days, giving up all other work.

I strolled back to the house and slept well that night.

At sunrise I set off again, past spreading, quiet mountains and valleys. I began to sing "Marching to Pretoria." Sana and Sayuma hummed along. I saw a patient, Tanawa, who had a brother, Magane, who had died of kuru the year before. Magane had been healthy for twenty-six years until developing symptoms of kuru and dying one year later of the disease. Only after his death did his sister, Tanawa, begin to notice her own unsteadiness. Since the end of the last rainy season, she had slowly gotten worse, able to walk only with a stick, and now not at all. This, then, could be a possible cluster of the type suspected but never yet found, showing that the incubation period did not randomly take over two decades. Rather, a very specific process was going on during this time. But as I gathered more information, it seemed the two siblings attended several feasts—over several years. They might not have been infected at the same time. I tried to ask and follow in *tok ples* simple questions—what was her mother's name? What was the next child's name? Everyone present laughed at my accent and mistakes—particularly a friendly older man. They had all felt lost in my culture; my uncertainty in theirs made us more equal.

On the trail back, I started to sing, "Working it Out," which I translated into Pidgin. Sana and Sayuma again hummed along. I started to become one with them, to trust myself and them. Formerly difficult, very steep portions of the trail became easier.

"What do you want most out of life?" I tried asking Sana and Sayuma. "What do you dream about?"

"Going to America," Sana said.

"How come?"

He paused.

". . . To see the medicines Carleton is making against kuru."

"But he doesn't have any yet."

"Then I want to see what Carleton is trying."

Sayuma butted in. "I have only two worries," he said. "I want Carleton to take my son Binabi to America, and then I want to visit Binabi there. And I want the Institute to buy me a truck to get around to see patients."

"America number one place," Sana repeated. They saw America as Oz—a land of wonder where everything was possible.

"But America has its problems, too," I said. They didn't understand. "People are unhappy in America for many reasons. There are too many people, and poor people." But these problems meant nothing here.

We passed a glen of particularly beautiful, dense, dark green trees. I stopped and looked.

"It is place bilong *maselai*," Sana told me. "A lake lies just inside the woods."

"Let's go see it!" I said. Sana drew back, panicked.

"You cannot go."

"Why not?"

Sana and Sayuma quickly looked at each other, at a loss for words.

"Because there are snakes there!" Sayuma finally said.

"Snakes?"

"Snakes . . . and *maselai*," Sana quietly added.

They believed in *maselai* (spirits) like "ghosts," haunting the woods. New Guineans believe that *maselai*—spirits of families, of ancestors, of particular places—determine how one fares in the world. How many yams grow in one's garden, whether one gets stuck in a rainstorm when one is inside or outside one's hut result from how good one's spirits are to one.

I began to wonder what functions these *maselai* served here. How do they relieve anxieties, or explain the world? How is this system of explanation replaced by science and religion in our society? How do people here look at nature, myth, good and evil, history, time, the future, the body, the universe, the mind and dreams? What exactly does civilization offer that these individuals do not have? I realized how biased the term 'civilization' was—suggesting a vast chasm between 'civilized' and so-called 'uncivilized' human beings. I was less sure now what the differences actually were.

The next day I hiked to a village even more primitive than the last—the huts were more dilapidated and crudely made, the wood more roughly hewn. The inhabitants owned few Western clothes. Some of the men wore native dress—merely old leaves gathered around the waist. Nonetheless, clothing was important here, its absence taboo. In many areas of New Guinea, the men wore only penile sheaths—long, hollow crypts into which the penis is inserted. But without these coverings, the men would consider themselves embarrassingly nude. Here, further from the road, the villagers seemed kinder and friendlier than elsewhere, less jaded by contact with Westerners.

The wide variety of physical features in this small population surprised me. Everyone, though closely related, looked different—even brothers or sisters. Genes clearly mix to produce an incredibly wide range of facial features. Such genetic mixing in such a small society, producing a wide variety of physical and probably personality features, confers diversity and hence, no doubt, evolutionary advantage.

The patient here also had had several female relatives die of kuru, but only one in her lifetime, ten years ago.

That night, I was exhausted. I washed up in the small sink in the back, then went to bed. The narrow cot was mushy, but I was glad at last to lay flat. Still, I couldn't sleep. My mind raced with images of forests and huts and patients. I propped a pillow over my eyes, trying to calm myself. I thought of Dorothy in *The Wizard of Oz*, leaving home and voyaging to a strange land, surrounded by unknown people and risking death, wizards, lizards, and other dangers. I began to doze off, but suddenly woke up. The whole house was shaking—trembling back and forth, as if one of the wooden posts holding up the structure was being sawed down. It was far more violent than the earthquake I had felt at Michael's. Alarmed, I quickly put on a pair of shorts and hurried outside to see what was going on. The building stood, supported three or four feet above the mud on four wooden 4" by 4" columns. A hairy black pig (part wild boar) was rubbing itself back and forth against the edge of one of these vertical beams, scratching itself and causing the whole house to vibrate. I turned around and tried to go back to bed, though the heaving and hawing continued.

Yet I couldn't sleep. I itched. I scratched my arm, my leg, my scalp. I felt a bite. I turned on my flashlight, looked on the sheet and saw a tiny speck jump—an insect, a flea. In horror, I stood up, pulled back the blanket, and saw two more dark-black bugs. I left the room, boiled a kettle of water, poured its contents into a bucket, added cold water, and hoisted the pail onto the ceiling hook in the shower room. Then I stripped, stood beneath it, and tilted the contraption to soak myself. I lathered with soap and then rinsed off. I changed the sheets and went back to sleep, furious.

The next day I trekked to a more distant village, Kanigatasa. The day promised to be long and hot. I hadn't been much of an athlete and had to push myself. But I was learning that if I persisted, I could do it. By 11 a.m., I was sweating profusely. Here, close to the equator, the sun squeezed down hard directly overhead. We passed sheaves of six-foot-high grass, now dried and white. My eyes concentrated on the ground six feet down and a little forward, watching every step, never looking up. I felt strained. Why was I doing this? For science, I told myself. I hoped it would be worth it—that I'd find something worthwhile.

Kanigatasa sat by the side of a small stream gushing over a water-fall. The hamlet probably hadn't moved for centuries, but I would never have known it. This place lacked any man-made artifacts or written evidence of history, of time passing. The past and present blurred. There were no indications of how people in the past lived differently, if they did. Here, the past seemed frozen, eternal. People indeed lived almost as they had since the earliest days of mankind's presence on earth.

The patient there, Salutu, didn't look to have kuru. It wasn't always easy to tell. In fact, I would later hear, when Stan Prusiner had come here a few years before, he wasn't permitted to see a patient who Sayuma said had kuru. Stan saw her at a *mumu* as she ran away. Based on this viewing of her, he diagnosed her as disease-free. She then deteriorated neurologically and died a year later. Diagnosis had to be carefully done and could easily be wrong. Another patient Stan had seen then went "to the bush" to get treated by a local witch doctor. She blamed her disease on Stan's having called out and written down her name.

I went over to Salutu now and asked her to walk. She did. I asked her to stand on one foot, then to put her hand to her nose, and finally to bring her index fingers together—all tests of coordination. She found these tests ridiculous, as of course they would be to someone who does not have kuru and knows that he or she can perform them perfectly well. To someone with kuru, these tests made sense and he or she would do them unquestioningly, knowing that physical ailments and limitations were about to be revealed.

Sana told me that another kuru patient, Saroma, was also here in Kanigatasa, visiting, though she was from Kalu. I wanted to see her, but the old men present decided it was best if I went and asked her husband first.

Kalu was several hours' hike in another direction. It would be a long day of hiking to go there and back to speak to her husband and see her, instead of examining her now. But I thought it best to respect native wishes.

That afternoon, we trekked back in the rain and mud. Luckily, before I had left New York I had bought a navy blue vinyl poncho at a sporting goods store. I now took the small parcel out of my knapsack, unfolded it, and slipped it on. It had a protective hood. The Fore had no raingear at all, and walked through downpours, soaked. My guides were amazed, however, never having seen a poncho before. They were surprised that a compact bundle could make such a practical garment.

Still, mud covered my feet, arms, and face. I thought of myself as in some army, straggling back from battle across vast distances. Finally, I arrived home. One man in the village, Busakara, saw me with the poncho on and came out to finger it in amazement as well. On the front porch, a man and his four kids besieged me, waiting for Roger, their eyes hungry, their tongues practically hanging out of their mouths. I felt obliged to invite them in, and offer them drinks and a snack. I cut and placed on the table two cucumbers—quickly devoured. Sayuma suddenly walked in and sat down, took in the situation, stood up, and walked back out, though leaving behind his bag—purposefully, I thought. An hour later Roger returned and talked to the man, for whom Roger was building a washhouse.

At the end of the day, I felt drained from hiking and trying to follow and talk in Pidgin. That night in bed I itched again and saw two fleas. I soon realized that every time I went into a native hut, I could catch fleas, and would have to shower afterward. That meant that the natives were probably all flea-infested. How did they manage to sleep?

I showered and returned to bed. That night, I dreamt I was on a yacht leaving the harbor of a large, glittering city at twilight. The boat was full of people my age, standing up—the men in navy blue sport jackets and white pants, the women in black dresses, drinks in our hands. The water was vast, quiet, and calm as we sailed away.

The next day my guides and I hiked to an even more remote village, Mentilasa. I sat down on the ground—there were no chairs, stools, or even logs on which to sit—while Sayuma talked with the local men. Near me, a young boy lay completely naked, pulling a dirty stick through his mouth. At first I thought of all the germs he was getting. Then I realized that he had probably already been exposed to every parasite and germ in the environment. Babies shat on the ground, wearing no diapers, underwear, or pants. One toddler slid around on his rear end over the muddy ground, littered with decaying matter and frequented by chickens. Several pigs tore apart the remaining tattered shreds of another pig they had attacked and eaten. Sayuma told me that pigs attack others' offspring. What an unhealthy environment and way of life, I thought. Civilization was probably overly concerned with health and cleanliness (for example, to conserve water, we could use shower runoff in toilets; we run from death and disease). But this other world harbored danger.

Sayuma told me that the patient we had traveled to see was inside a hut. Sayuma talked longer to the men of the hamlet. He then told me that he had been wrong and that the woman was not here. I suspected he was lying, and debated just going in to see her. But as it was my first time here, I deferred, though I'm sure Carleton would have just gone in. Sana then told me that the discussion with the men had concluded that I should get the woman's husband's permission first. I thanked Sana for having told me the truth.

Afterward, I told Sana and Sayuma that I wanted the truth and would accept the consequences of the truth, but not lies. Sana understood. Sayuma didn't pay any attention. I tried to be firm and clear as I'm sure Carleton would have. Even in this very different culture, first impressions are important; some people one likes and trusts right away. Others less so.

Fore landscape: dawn.

Fore landscape: morning.

The author in the Highlands.

The Highlands.

Fore huts.

A new style of hut, seen on the author's most recent (1997) trip.

Trekking.

Fore fences.

A female kuru patient.

A male kuru patient.

Older villager informants.

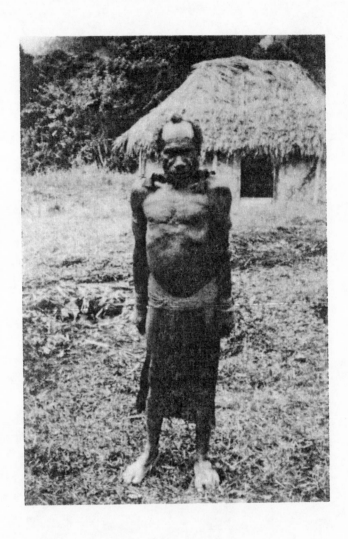

Residents of Fore hamlets.

A Highlands pig (eating another pig).

Carrying loads on the road.

Carrying rice on the road.

What Do I Have to Do
to Be Saved?

"You haven't met Jake and Mildred Lewis, have you?" Maryanne asked me that night, referring to the only other Caucasians for many miles around—missionaries.

"No," I said, not particularly interested.

"You should. They're Americans, too. I'll introduce you." I was still wary, not sure what I would talk with them about. Despite the airplane trip, when I thought of missionaries, I still imagined monks in long frocks venturing to South America in the 1700s. I didn't even know Protestant missionaries existed—only Catholic ones, and I only knew that from movies and history books. Before coming to this country, I had certainly never met any, let alone socialized with them. But Maryanne had become friends with them—as allies here in the bush.

I agreed to go with her, and one morning a few days later, we walked down the road and came to a wire fence—the first I had seen here. It surrounded their home. We opened the gate and walked in past a few chickens clucking about the yard. Maryanne rang the doorbell— probably the only one for hundreds of miles. A tall, thin man with white hair and gold-rimmed glasses greeted us at the door. "Oh, hello," Maryanne said shyly. "This is Bob. This is Jake Lewis." Maryanne looked down at the ground. She had grown up in rural Australia and was not used to social dealings. He escorted us into the living room, furnished with white doilies, potted plants, and captain's-style den furniture looking like it had been imported directly from an Ethan Allen showroom. Mildred Lewis came in carrying a tray of lemonade. I sat down on a rocking chair beside an exercise bike that I later learned was

Mildred's and that had a view of the mountains through the frilly lacey curtains hanging on the window. The room stood as an outpost of the Midwest here in the tropics.

The Lewises had come to New Guinea ten years ago. He was from South Dakota, she was originally from North Dakota. She was short and, in a light-blue plaid gingham dress with a white collar and belt, looked as if she were still somewhere in Fargo. The corners of her eyeglasses pointed outward, bejeweled with tiny rhinestones— out of fashion since the late 1950s or early 1960s, though recently funky on the Lower East Side of New York. The Lewises seemed to live in a time warp.

They invited us to stay for lunch. We sat down and they said grace. Afterward we drank a fragrant cinnamon, orange, and spice tea around their coffee table. She asked very perceptive questions about Carleton and kuru, then asked awkwardly about other topics. "What school did you go to . . . What does your father do? Do you have any brothers or sisters? Are they married?"

The Lewises had a very detailed understanding of native culture, and spoke *tok ples* fluently, yet had no interest in the local culture for its own sake. They kept abreast of all the neighborhood gossip and scandal—how men had lost their stores' contents or sons' school fees, playing cards—as evidence of the heathen's evil ways.

I realized the wide difference between the two groups in this country—natives and expats. Each moved in its own separate track. I felt somewhat in the middle, an observer. What a shame that usually only missionaries spoke the local language—to convert the people. If anthropologists came and learned a language, they were by now long gone, usually staying for one to three years and then leaving for good, perhaps to return briefly decades later.

"We just had another chicken stolen," Jake Lewis told me. "We're sure it's Sayuma's son, Jason. We've seen him snooping around here at night, and we caught him once before. He's a real rascal. You have to be careful with him."

Jake owned a general store back in America and remained a small business man at heart here. "We use the natives to make souvenirs in the school—beads and arrows with feathers. Then I have these shipped back to the States and sold through the church to raise money. You'd be surprised how much they bring in." He showed me samples. Yet these trinkets had never been made in this culture before. Mr. Lewis had taught the Fore how. These objects conformed to his idea of what a primitive tribe would construct.

When he found out I would be going to Hong Kong when I eventually left New Guinea, he gave me names of stores for buying inexpensive camera equipment and suits, and tips on how to work the duty-free shops. "Buy camera lenses, ivory, and jade. They're all cheap. Ivory will be a lot scarcer in the future, you know. Shipping goods home is a problem, though. It takes months, so go with an empty suitcase." I would have found all this conversation somewhat boring, if not repulsive, in the States. Yet here it was almost welcome—just to sit in a cozy den drinking tea all afternoon, talking with Americans. I was surprised that this was important to me—that I missed home that much.

I got up to go to the bathroom. On the top of their toilet seat, beneath a fake baroque mirror, stood a tiny red, white, and blue book. The cover read, "What Do I Have to Do to Be Saved?" The answer inside was to convert heathen to Christianity.

I later found that a few years earlier, the Lewises' son had died in a motorcycle accident here in the Highlands. Mildred and Jake were devastated, but then decided that it was God's will that they remain in PNG and devote themselves even further to their work. Nonetheless, they complained bitterly and constantly about it.

Still, they taught me a lot and it was helpful hearing their difficulties here. We commiserated. Primitive cultures make strange bedfellows.

A few days later, after two weeks in Waisa, I decided to visit Goroka. I had not yet met with Michael, whom I assumed would be there now. I also remembered seeing at the NIH a computer printout of all kuru patients recorded since 1957. Government patrols had compiled names in village census books and these had been added as well. I could have the printout airmailed to me. I asked Sayuma to travel to Kalu in my absence to ask Saroma's husband if I could see her on my return. Sayuma said he would.

I took a PMV (or Private Motor Vehicle—usually a pickup truck—which gave people lifts for a fee). I had journeyed along this knotted road once before, but now everything looked different. The first time, I had searched every hut, the look on every face. Every detail stamped itself on my memory. I had peered behind trees to look into valleys for scenes of a different world. This time, however, I didn't scan as before. Instead, I observed the gradual change in landscape from mountains to wider valleys and from primitive villages to the town.

I now returned to where I had started in the Highlands. In the two-week interim, I had survived in this new and unusual realm. Only now in retrospect, out of the situation, did I recognize the full weight of unarticulated worries that had lurked in the back of my mind, making me vaguely nervous. I had overcome fears and uncertainties about seeing kuru patients, and endured in this primitive world. I had experienced the *rites de passage* of entering a different culture, journeying from civilization to essentially the Stone Age. Clearly the Fore were now emerging from that era. Villagers now played cards, rode on trucks, and wore bits of Western clothes. Yet in many regards, these changes seemed superficial. Again and again I was impressed by how much had remained basically the same. The changes were outward; as I would soon see, beliefs about sorcery, spirits, and cargo cults all prevailed. Over upcoming decades these beliefs would, I suspected, alter. But for now, the Fore suggested much about how, prehistorically, human beings lived, organized, and viewed their lives. New observations and questions now filled my mind. I returned to Goroka a somewhat different person.

By this point, I had learned quite a bit of Pidgin. Ami was most impressed and complimented me. I arrived at the Institute to find out that Michael was delayed returning, and would not be back until the next day. I was nervous, embarrassed by my lack of training. But what would be proper training? I had gotten by in the bush and already seen and learned much.

Still, I was so nervous that I almost wished that he wouldn't come, and that I could return to the village and see him in the future. It would be foolish, however, to run away. I forced myself to think of the questions I had for him: How detailed an examination of patients should I make? How extended a pedigree should I collect?

Mail had arrived for me—mostly from friends. One letter was from my sister, who had gone with my mother to the local public library and looked up New Guinea in *National Geographic*. "If I knew how primitive and undeveloped it was," my mother had told my sister, "I wouldn't have let him go." But I was already here, and would have gone nonetheless. I was now an adult.

I went to the bank to exchange money. Six employees sat behind a wooden partition, doing nothing. One woman took my traveler's check and stamped it, then gave it to a man who wrote something down and gave the packet to a third person, who acted as a teller. The bank sold T-shirts: "Papua New Guinea Banking Corporation," the front read, beside two carved masks. The back said, "Nambawan Haus Moni Bilong NiuGini"—"The best bank in New Guinea." I needed more T-shirts, and bought one.

I found one snack bar in town—two tables with chairs under thatched umbrellas. I bought a Coca Cola and a meat pie and sat down. Another Caucasian walked down the street. He looked a little older than me, but smiled when he saw me, relieved to find a compatriot, and came over.

"Mind if I sit down?" he asked in an Australian accent.

"Not at all."

He was here on vacation. "I always wanted to come here," he said, "but it's been hard getting around or seeing anything really exotic. What are you really doing here?"

"Research," I said. He asked all about it.

"They were really cannibals?" he asked. I began to explain. As I talked, he suddenly took out his camera and snapped my picture. He obviously saw me as a part of his adventures here in this bizarre land— a freak to include back home in a slide show of his trip. I quickly finished my lunch and left.

On my second day, Michael arrived. He was a tall man with a long brown beard, and wore a work shirt with two large breast pockets filled with pens. "Tea?" he asked as I sat down.

"Please," I said.

"What have you been up to out there in the bush?"

"Trekking to see patients with Sana and Sayuma."

"Good," he said, smiling for the first time. "There's another guide, too—Ganara, out on the fringe of the populated area in Paigatasa, a very primitive and remote region."

"Should I go to visit him, or send him word that I'm there in Waisa?"

"Oh, he'll hear. He'll come to you in due time." Michael was reassuring, and I now felt more on top of what I was doing.

"What about patients whose kuru is questionable?" I asked.

"Time will tell. We'll wait and see what happens." He looked down at his desk and restarted his work. The conversation was over.

I was back at Michael's house when the phone rang. There were no answering machines in New Guinea. I picked up the receiver. It was Ami. "Your friends are here," he said.

"My friends?"

"Yes."

"Who are they?"

"I don't know."

"What are their names?"

"They didn't say."

I decided to head over. When I walked in the Institute, to my surprise sitting in the hall were the Whittiers, whom I had met on the plane three weeks earlier—though it seemed like months.

"We were driving by," Mr. Whittier said, "and saw the sign saying Institute of Medical Research. We remembered that that's where you were working, so we stopped by." It was refreshing to see them—Americans who weren't missionaries.

"Our trip is over," Mrs. Whittier added, "we're now heading back to the States. We'll be home in five days."

I offered them tea, now knowing where it was kept and I suppose feeling somewhat at home here. "How was the trip?" I asked.

"Great. We saw Walter in Mount Hagen. He returned to the States last week. And we ran into Mark and Steve on the coast. They're scheduled to be back in San Francisco today. They had bumped into Sandra and found out she was on the same return flight as they are." I felt I would soon be the only one of the group left here.

"The idea of going home sounds nice," I said. Home seemed worlds, universes away.

"But you have such an exciting year ahead of you. You'll look back on it very glad you had done it."

I sighed. It was true. I was surprised that I had managed to lose sight of that. But being here was hard, given the hassles of traveling or getting things done. And the full significance of this country hadn't hit me. "Do you want us to mail anything or call anyone back in the States for you?" Mrs. Whittier asked.

"Maybe my parents," I said.

"Sure, give us the number." I did.

"We're glad to be going back," one of their daughters confessed, rolling her eyes. "We've traveled all over Africa, South America, and Asia. But this has been the most difficult place to get around in. And there's nothing to see here." It wasn't set up for the tourists.

Eventually, it was time for them to go. We stood up and I walked them to their landrover. They piled in, closed the doors, and started down the driveway. I stood at the top of the dirt hill and waved goodbye.

Sadly, I returned to Michael's house. He came home for dinner, then went back to work. I went to sleep at midnight. He came in after that. In the morning when I rose—at 7:30—he had already risen, showered, breakfasted, and returned to the Institute.

"Are there any newspapers here?" I asked him later when I happened to see him back at his office.

"Only in Pidgin."

"Which do you read?"

"I don't."

"Don't you miss the news?"

"No."

"You have a lot of good books around," I said, trying to make conversation.

"I like to read." He was a gentleman with English reserve.

"I see you particularly have a lot of Proust."

"He's my favorite author."

"I've never read him."

"You should." Back at Michael's house, I started Proust, interested partly in what drew Alpers this strongly. Proust, I soon saw, sought the timeless in art and memory. I thought of Ezra Pound's line that literature was news that stayed news.

Here, time was suspended, as if the distractions of centuries hadn't occurred. Carleton and Michael once sat on a mountain top and agreed that they would write only those papers that would still be read in 50 years. Similarly, Michael preferred to read books written at least 50 years ago, and his musical collection was mostly renaissance and baroque. The only music he had that had been written in the past 100 years was a Joan Armatrading album his daughter at Oxford had given him for Christmas. I played it several times, having grown up with folk rock. The songs' cool rhythms reminded me of the world I had left behind.

I then put on Mozart violin concertos, and made a pot of Lapsang Souchong tea. Michael had a cabinet full of the over two dozen varieties of Twining's and Fortnum and Mason teas in tin containers that can be stored here in the tropics. The teas themselves came from all over the world—Russian Caravan, Darjeeling, Jasmine, English Breakfast tea, Irish Breakfast tea. Just seeing the names on the orange, purple, and green boxes filled my mind with images of the rest of the world. From the one bakery in town I had bought fresh bread, and from the Steamships store, fresh butter imported from New Zealand, and English marmalade.

Outside, delicate pine trees floated against the pure blue sky to Mozart and Bach. Waves of clouds tumbled, splashing mist quietly across green mountains. The music stopped. From atop a hill, far off in the bush, birds cried. Trees rustled like the sea.

I realized how much Michael, too, sought the timeless in music, nature, science, and culture. More than anyone I had ever met, he had tried to transcend time by using it effectively and thus conquering its restraints. He had first studied mathematics at Oxford and then arts. He collected the most primitive native shields, most no longer made. They were raw, original, and pure. Here in New Guinea, was man—universal—unaffected by time. There was something essential here. The force of nature operated strongly.

Over the next few months, when frustrated, I tried to adopt Michael's stance—to stand above the fray. But it was hard. Perhaps it was merely that he had lived more. But he taught me to try to stand back, and be wary of hype.

That evening, Michael had older visitors—English physicians—who talked about the CIA. I tried to comment, but felt out of place—American and young. I had never seen myself as particularly American, and in fact was critical of the United States in some ways. But for the first time I was now consistently seen as White and as American. Others' views of us are shaped by the culture we're in. I was learning a lot about social roles. I would also be glad to go back to my own self-created world at Waisa with my own room and belongings.

The next morning I went to the Institute and mailed letters. I toured around and happened to find on a cluttered table in the back an old beat-up manual typewriter, New Guinean by way of Australia. It wasn't standardized, and had a British pound sign—instead of an American dollar sign—which I'd never used before. It also didn't work well, and had a faded neocolonial air about it of tropical undergrowth and lassitude; it was stained with spilled drops of white paint, and had a note taped on, the Scotch tape sloppily applied, now yellowed and barely adhering:

"VERY IMPORTANT NOTE: DO NOT STRIKE THIS INNOCENT MACHINE: DO NOT TOUCH IT IF ANYTHING GOES WRONG: PLEASE CALL REMINGTON MAN: A. TALERATA."

A. Talerata was Ami. But I decided I wanted it. I could keep field notes, as well as observations and my sanity; remember and use English; and organize the experience to see a higher purpose beyond the logistical difficulties that consumed much of my day. I asked Michael, who said okay. Later that day, I took it with me on a PMV back to Waisa.

It was relaxing, returning to what I now felt was home. Maryanne was planting a small garden on the side of the house. I helped her as the sun began to slip down. I weeded around a passion fruit plant growing along the front of the porch. The pressures of Goroka faded behind me, here in this calmer life in the mountains. I felt free, and was surprised that I now felt Waisa was my home.

I reflected on my life here and to date, and realized I had done okay. I had people back in the States who I loved and missed—more friends than I thought or realized. I set up the typewriter on the small desk below the window in my room, and began to write about my experiences, trying to describe my confusion and wonder. I was glad to have the typewriter with me.

That evening, I cooked crepes, having bought ham and Swiss cheese in Goroka.

"Sounds like Sayuma's been busy in your absence," Roger told me. "He told us he went to Kalu yesterday to arrange for you to see a patient named Saroma there this week."

Butter

That night I took a hot shower, dried off, then wrapped myself in thick, warm blankets and plopped down on my soft pillow and bed. The rain splattered on the roof, on the vastness of valleys and mountains and on all houses everywhere, it seemed. I listened, protected, to the squall's symphony. Early in the morning I woke; from 5 to 7 a.m. the rain still fell as I lay warm and bundled up, half asleep. At last I had peace and rest from exhausting drives, treks, and discussions.

In the morning there was a knock at the door. Sayuma's son, Jason, who lived with his father in Karamuni hamlet, had come by to sell lemons and avocados. He had picked them by climbing trees that grew on our property behind the house. Some Westerners had apparently brought lemons and avocados out here a few years ago, and thrown the pits out in the backyard. In this fertile soil and climate the seeds had sprouted. As a boy in Manhattan, whenever I planted citrus seeds and avocado pits on the windowsill, the plants never grew more than six or eight inches, or produced more than a dozen or so leaves—let alone a flower. Yet here in PNG the lemon seed and avocado pits had matured to full trees, flowering and fruiting.

The natives did not eat these fruits, which were not part of their traditional diet, and had never been formally introduced and encouraged by Western missionaries. Both these fruits were seen as strange anomalies—something only Westerners ate. We are all caught in our respective diets.

The natives called avocado by its name in Pidgin—"butter"—the word applied because of the consistency. I never saw the natives eat real butter, or any evidence that they knew what it was, but that didn't seem to matter. Words become accepted regardless of their roots. After all, even for children, words initially are merely sounds, arbitrary in and of themselves, adopted because of the language into which the child is born. In a different culture, the same sounds could have entirely different meanings. A child learns the roots of words only much later, if ever.

Jason also sold us the large eggs of a cassowary, an ostrich-like bird that had grown increasingly rare since the building of the road and the appearance of trucks. "Is it fertilized?" Maryanne asked.

"Oh, no," he assured her. We wanted eggs. Otherwise, our own protein intake usually consisted of canned tuna and Spam that we brought out from Goroka once every month or so. We would buy eggs and cheese there as well, but without refrigeration, these lasted only a few days in the bush.

After he left, Maryanne cracked the egg open. A tiny red heart beat inside it with two tiny crimson arteries branching out into the yolk. A chick was beginning to form. The egg had been fertilized. I felt terrible about it. "Jason must not have known it was fertilized," Maryanne said.

"But how would he know one way or the other?" I asked.

"Oh, the natives are *very smart* about these things," she said with wide eyes, as if she believed they possessed almost natural powers of divination. I was dubious. The following week, she fell for it again when someone else sold her an egg. "Are you *sure* it's not fertilized?" she asked the man.

"Oh no," he answered.

When she cracked it open, another heart beat.

"He must not have known it," Maryanne said innocently again.

I remained more suspect. He just wanted money.

Sayuma came by the next day. "I want to go to America," he announced.

"Yes, I know," I replied.

"I heard they have new teeth they give people."

I was confused. Then, looking at him, with his two remaining teeth, I understood. "You mean dentures." Sayuma, who didn't know or understand the word, nonetheless recognized that I realized what he was talking about. He smiled widely, his two teeth showing, and nodded.

"Em stret," he said in Pidgin (from the English, "them straight," meaning, "that's right").

I went into the house, but Sayuma continued hanging around the porch. Roger and I were just sitting down to eat. Sayuma then came inside. "Is Sana here?" he asked as we sat eating.

"No," I said. Still, Sayuma stood there. I felt awkward. "Do you want to join us?" I asked, with reluctance and annoyance in my voice. We had the food, and I didn't want to lower morale. But I didn't like being forced into the situation. Sayuma immediately sat down, but ate only what he couldn't get at his place—the Western foods I served— spaghetti, and the entire basket of bread I set on the table. He drank all of his tea, but didn't touch the tomatoes, beans, or corn—all obtained from local gardens. He didn't even taste them.

The next morning Sana arrived at the door. "Last night I had a dream," he said.

"Really?" I asked, fascinated.

"Yes. You and I were drinking beer together. We both had gardens. Other people surrounded us. I got drunk and asked you if you had gotten my passport. You said yes. I asked you if you had bought the airplane, and you said yes—a jet. Then you and I were sitting in the jet."

I got up and walked to the kitchen to get some coffee. Maryanne was in the kitchen. "You wouldn't believe what Sana told me," I said to her. I told her the dream.

"But New Guineans believe that a dream is a sure sign that something will come true," she said, nodding, seeming to believe this, too. In fact I was hoping to travel west when I eventually left here, continuing in the direction in which I had come and visiting Asia on the way home.

Maryanne attended services at the Lewises' church every Sunday. Roger had asked if I would volunteer taking turns with him manning a small library across the road—a building, once an Aide post (a small structure where a visiting nurse, doctor, or trained medical assistant had periodically stopped to offer medical care). Michael had then stocked the structure with old books and magazines. I gladly agreed. On my shift, two children and one woman wandered by and shyly chatted. I showed them books with pictures of plants and animals found outside New Guinea. The visitors looked, but didn't seem very impressed by what I presented.

The following Sunday, Maryanne asked me if I was interested in accompanying her to church while Roger sat at the library. I decided to go. Though Jewish, I had visited numerous churches in Europe. Why not one in New Guinea? The bamboo structure stood on a rise across from the Lewises' house. We followed village women to the service, their bare feet on the ground, their toes swimming through the soggy mud, painting their black feet with the oranges and reds of the clay soil. Straps rising up and around their foreheads held *bilums*—bags woven of native fibers and colored with faded natural dyes. The satchels hung down the women's backs, tugged down by the weight of muddy potatoes and beans. The women waddled over the matted floors and huddled close together. One infant cried. His mother raised her hand under a shirt, pulled out a long breast, and pushed the tip to his mouth. Other babies lolled about their mothers' sagging breasts as well. The women sang, their soft voices whispering like the wind. They clapped their hands—used to toiling in gardens—softly, separately, like the patter of rain on a thatched roof.

The men strode in late with pride, and stood in the back singly. All the local native women were there. The only men present were the paid church "regulars"—the Lewises' drivers and "houseboys." Three times as many women as men filled the church. There were no chairs or pews—instead we sat on the ground on bamboo mats. Mr. Lewis read from the Bible about, "Nambawan *maselai*—Jesus Christ." He read a prayer, "Papa bilong mipela" (Pidgin for "Our father," literally "father of us," as "mipela," from the English, "my fellow," means "my group"). "I stop long heven" ("he stops, or is, in heaven"). "Nem bilong em i stop i holi" ("his name is holy"). "Kingdom bilong em i mus i kam."

While the congregation around me recited their prayers, the only ones of my own religion, Judaism, that I was able to recall while sitting there were fragments in English, which I now mumbled—to protect myself and feel more comfortable.

Afterward, the women sat on the ground in front of the church. I realized this was their only time to be together socially during the week. The men could play cards—a leftover from the days of the men's houses, when the group's sole religion was a male-only affair. Missionaries helped end the men's houses and the warfare, and gave the women a social institution of their own. Yet I now began to see how the natives heard missionaries preach about Jesus Christ, as "Nambawan *maselai.*" The natives, seeing these Westerners with airplanes, trucks, tinned food, and radios, deduced that this *maselai* was good. New Guineans joining the Church were rewarded with goods—clothes and food donated by parishioners back in the United States or elsewhere. The natives, I would later see, had further concluded that these were good spirits. The native view incorporated the Western view more than vice versa.

Miss Hazel McGill, an elderly missionary colleague of the Lewises, was strolling from one circle of people to another. The natives seemed genuinely to respect and admire her. I had heard about her from Carleton. "We should invite her to lunch sometime," I said to Maryanne.

"What a wonderful idea," Maryanne responded. "Why not today?"

We sauntered over. Maryanne introduced us. "We were wondering if you might join us for lunch," I told Hazel.

"Why, I've been invited out to lunch," she happily announced to a Fore woman standing next to her, who I don't think understood.

At noon, Hazel, accompanied by Mildred Lewis, arrived at our hut for a lunch that Maryanne mostly put together. Maryanne had never before had the missionaries over, and she seemed pleased to have the chance.

"Why, this is so nice," Hazel announced as she sat down at our small table.

"How long have you been in New Guinea?" I asked her.

"Seventeen years. I came when I was 65." Her father had been a preacher, she told us, and she was born in Pittsburgh—"a dirty city"— where she went to college and then became a high school teacher. She received a masters degree at a religious university and subsequently found her "calling." Her mission work took her throughout the United States, but mostly to New York City, where she stayed for twenty-five years. She fed and cared for alcoholics in Times Square and the Bowery.

"Was that one of your more trying experiences?" I asked.

"Oh no."

"How did you end up coming to New Guinea?"

"When Mr. James—the first missionary in this area of the Highlands—arrived back in New York in the early 1960s, after three years in New Guinea, a mutual friend called and told me I had to hear a

wonderful mission story. So I went. In their first three years, the James'
had brought the Gospel to two thousand souls. For days after I heard
James speak, I told friends about the important work he and his wife
were doing. At some point, I realized that the Lord was to bring me to
New Guinea. New Guineans had no religion, so did not have to un-
learn anything, as those in other countries did." (But, I thought to myself, the New Guineans *did* have their own religious beliefs.)

"I went to all the shipping companies in New York," Hazel continued, "trying to find out how to get here. No ships traveled from the
US to New Guinea. I would have to go to Australia and see what I
could find there. Four weeks after I heard Mr. James, I set sail." At 65,
she was already at an age when most people look toward retirement,
not a new job—let alone a new life in a new and unknown country.
"When I arrived in Sydney it was winter, and cold. I knew no one. I had
no idea whether I'd be able to buy supplies or furniture once I got here,
so I bought things in Sydney. Two weeks later, I flew to Lae and then to
Kainantu. I knew no one there, but eventually found my way to Okapa
and located the *kiap*," or local policeman. "He said I could not go. 'If
you get killed,' he told me, 'the government can take no responsibility.'
But I wasn't going to let him stand in my way."

As she spoke, white light diffused softly over the pale wrinkles of
her forehead and her white hair. She had an ethereal, almost timeless air
about her. "The road," she continued, "was even poorer then than it is
now. One of the bridges was washed out. Teams of natives had to pull
my car out. That was my first direct contact with New Guineans. Finally, I arrived at the James' mission, at the very end of the road." I had
seen the James' compound, still standing in Purosa on a hill, surrounded
by a tall fence. "I had thought I would be stationed there with them,"
Hazel continued. "But to my surprise, Mr. James said no. He had built a
house for me miles away in the jungle. James himself took me there.
When I arrived, there wasn't even a door. He warned me to install a
room of mosquito net to sleep in. Then he left me there with an inter-
preter and disappeared, saying he'd come back in several weeks."

"I stayed, and began to meet and convert the natives. Most had
never seen a White woman before, and tried to feel my breasts—to see
if they were real—and my body. Up until then the natives had assumed
there were only White men. That was all they had seen."

"Jesus Christ!" I muttered. Suddenly I felt embarrassed, sitting with missionaries and having taken the Lord's name in vain. But Hazel didn't seem to hear, and continued.

"A month later the *kiap* from Okapa arrived and told me I had to leave. 'You have no right to be here,' he said, 'since you do not own the land.' He threatened me as we sat in my small hut. All the while, I kept thinking of the line from Psalm 9: 'Let man not prevail.' This too shall pass."

We finished our main course and I got up to clear the table, assisted by Mrs. Lewis. "I hope the dessert's okay," I said to Mildred in the kitchen as I lifted it—a Jello and fruit salad I had made—to bring in.

"She'll say it's good," Mildred said, "Even if she doesn't think so." I was surprised to hear one missionary disparaging another.

Hazel spoke Latin, German, French, and Spanish, but said that Fore grammar is the most complicated and unusual. Sentences are structured object–subject–verb. "But none of the people could tell me this complex structure, though they all use it." She was surprised, but I wasn't. A person can perform a complex procedure without being aware of the rules governing his behavior. We all walk, but would have trouble explaining how.

"The local language has been mistranslated," Hazel added. "White men have made many mistakes trying to understand it—for instance with place names. The natives didn't always have place names. Fore place names now listed on government maps are all wrong. 'Asa' as a suffix means, 'is where I am from.' So that the names now given to villages, and appearing on maps—Mentilasa, Agakamatasa, Paigatasa— technically should be: Mentil, Agakama, Paigat. The Fore had answered with correct grammar when posed questions, but they were misunderstood." (I realized, however, that their language paralleled others—with the use of "de" in French, "von" in German, and "van" in Dutch.)

Hazel continued, "When the Australian patrolmen asked the Fore's neighbors, 'Who lives there?' The answer was 'The people who live in the grasslands down below,' or 'Fore.' It wasn't the name the Fore as a group used for themselves, yet it now became their name. The Fore, in turn, did the same thing in naming yet another group. Only recently have the nearby Kukukuku people, at the strong encouragement of Westerners entering the area, adopted the name 'Anga.' The Kukukuku had no name for themselves. That name was used exclusively by their enemies, and considered an insult." The Kukukuku included 13 different language groups. The only words they had in common were sexual and derogatory terms and the word for "house"—*anga*. Carleton and a linguist suggested that the group use that as a name.

"These people are just so difficult to understand!" Mildred said.

"But you have to try," Hazel explained.

"I think you just have to watch yourself and be on guard with them at all times," Mildred continued, "A lot are rascals. They're constantly stealing chickens from us."

"They're probably hungry and poor," I said.

"But they have to learn to buy things," Mildred said. "How are we going to help them in the modern world if they don't learn that?" Hazel, patient, operated from genuine warmth, not just doctrine or ideology. New Guineans quickly frustrated those who worked based on abstract principles of Christian charity. In the face of vast cultural differences, flexibility was key. "Do you know that Sayuma's son, Jason, now steals chickens from us practically *every week!*" Mildred continued. "Jake is ready to kill him. The mission dog, Snoopy, used to be a good watch dog, but has gotten too old to be on the alert."

"No," Hazel said, shaking her head emphatically. "Snoopy hasn't gotten too old. He has just lost a sense of purpose since you've bought and are now training two new puppies, walking and feeding them three times a day. Snoopy refuses to eat, sleeps more, and walks slower. He feels purposeless and left out." (I thought he probably had worms.) "We all need the will to live," Hazel continued. I realized that what kept Hazel going was just this very will to live, her firm belief that she was needed to help people—a crusade.

"Tell me about your work," she said. I did. "What you're doing is so important," she reflected. "Kuru is such a terrible disease. Patients suffer so much. But those that have the Lord in their hearts do better. Tasa, Soba's sister-in-law, had kuru and died last month. One week before, she displayed no symptoms. She was talkative and pushed her blanket away from herself when I visited her. She had had kuru four years ago and stumbled around with a stick, but never stopped working in her garden. Then she got well. This time, though, she lost the will to live. Those that have the Lord in their hearts are helped. Those who do not know Jesus lose the will to live, and die."

"But kuru is caused by a virus, you know," I said cautiously.

"Oh yes, I know. But when they actually die, is when they give up. And if they have Jesus in their hearts, they do better. There is so much ignorance we have to fight—so many dark ways. Tasa simply did not have the Lord in her life. They also practice euthanasia," she continued. "Tanawa died a few days ago. I think they killed her. They simply turn patients on their stomachs. With weakened muscles, the patients can't breathe and suffocate." I realized that euthanasia is not merely a withholding of available treatment—none was available here—but a deeply human response. Here, then, was a cultural argument in support of euthanasia—its universality.

"It's such a shame they do that," Mildred said. "We teach them it's against the Lord's commandments."

"This whole kuru epidemic is so sad," Hazel said. "But they refused to give up cannibalism. Why even last year, I heard rumors about a case of cannibalism occurring near Paigatasa."

"By the way have you met Ronald Monroe, yet?" Hazel asked, referring to another missionary in the region.

"No."

"You should. You'll learn something."

"Really? What?"

"I can't really say. You'll have to see for yourself. But be careful. He puts something between himself and the Lord."

"What?"

She hesitated. "His ego."

By now it was 3 p.m. The afternoon had passed quickly. I wasn't used to sitting in people's houses for an afternoon. It was cozy and I enjoyed it, despite the conversation's theological drift.

Maryanne asked if Miss McGill would like to close our lunch with a prayer. "Thank you, oh Lord, our God Jesus," Hazel proclaimed with enthusiasm. "Help us Jesus, even in our times of weakness. Thank you for bringing us together with others who know your truth."

"Amen," Mildred added.

"Amen," I added to myself.

Fore mothers and children.

Removing lice.

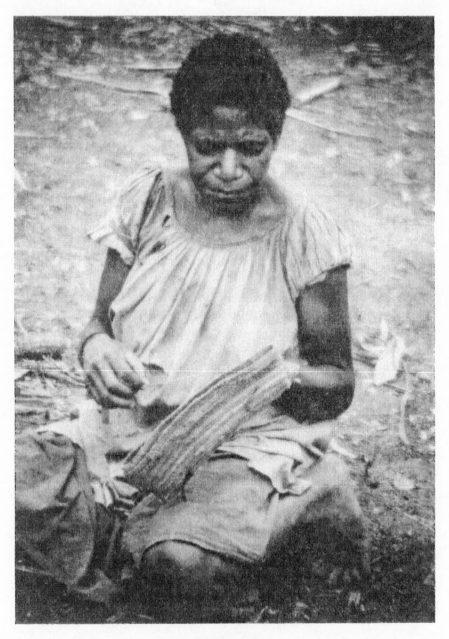

Weaving a bilum.

Fore children (one with a painted face).

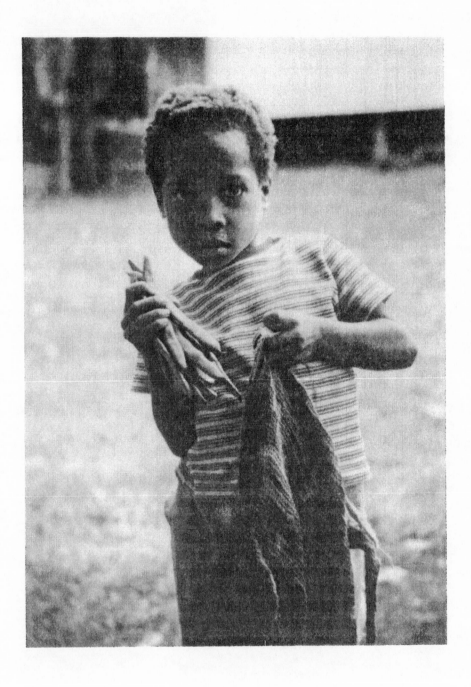

School yard—with children wearing only leaves around their waists.

Talk Place

Two days later, Sayuma told me that Wasoru had died Friday night. I was the only Westerner to have ever examined her. It was important for me to see all reported cases because the major evidence that kuru was spread via cannibalistic feasts—and did not, for instance, linger in the soil or spread genetically—was epidemiological, the fact that the disease was steadily decreasing, and that children and young adults no longer died of it as they once had.

The next morning, Sayuma said a council member would be driving from the road *bung* (junction) and could give us a lift to Mentilasa to see another patient. But when we got to the car, it was just parked without a driver. No one had any idea when it was going to go, if it was going at all. Thus, we walked. Along the way, I heard about two other patients from passers-by. In other words, Sayuma hadn't been doing his job well. He charged regularly for PMVs to go all over the place, but there were still some patients he didn't know of. He carried around a little hard-bound notebook for listing trips he had made looking for kuru patients. Michael had taught him to make notations of dates and the names of various villages. Sayuma could mimic the characters, copying them and thus potentially keeping track of his travels. But I had perused the list, and knew that many of the excursions never occurred—reported to have taken place on days when Sayuma had played cards all day in Waisa. In just the time that I knew of, Sayuma did not go to Mentilasa on Thursday or Friday as he claimed, or any other places shortly after I had arrived in January.

I wanted to see two patients in adjacent hamlets of Mentilasa. However, my guides wanted to detour elsewhere first to see another, recently deceased patient's family and stop off in Mentilasa on the way back. I acceded, but by the time we eventually arrived in Mentilasa, word had gotten out of our coming. The first hamlet was deserted. Finally, a boy appeared. He said the patient and the patient's wife were fixing a garden fence in the bush, and would be back later. I didn't believe it. We went on to the second hamlet, where we were told that the patient had gone away as well. I left word that I would return the next day.

We started to hike back to the first hamlet to check if the patient had returned. On the road, a PMV came rattling by. By now I was tired from eight hours of walking. I also wasn't sure at all that the patient would be back in the hamlet. I couldn't take people at their word here; I had to be careful all of the time, aware of their resistance. We hopped on the truck. On the way home, Sayuma only spoke to me to tell me that we needed an Institute car. He had brought along his brother, whose PMV ride I had to pay for as well. Sayuma wanted his brother to work for the Institute and no doubt wanted to be able to claim that the brother had helped me on the day's patrol. But I refused to be a pawn in his games. As we rode, I decided that if his brother tried to join us again tomorrow, I'd tell him he couldn't come.

The truck drove into mud at one point and we all had to get out and push it to the top of the hill, and then jump back on as the jeep started down. But the PMV trip cost just 50 toya—half a kina. That was a far cry from what Sayuma was charging the Institute—2 to 4 kina a ride, sometimes seven days a week. Kina, the national currency, were minted as round, donut-shaped coins with holes in them, since native groups had previously bartered pearly "kina shells," also bored with circular openings. A kina exchanged at U.S. $1.60.

As we neared Waisa, Jake Lewis pulled up and offered me a ride. I reached for my bag. "No," he said. "Let them take it back for you. Make them carry all the bags from now on and they will."

In the morning, I returned to Mentilasa, but the patients had again disappeared. I then traveled to a further village and interviewed three informants—two elderly men and an older woman. The men were prideful but ignorant, and came from previously warring villages. All disagreed with one another. I asked questions that should have had one answer and got two or three contradictory responses.

"Who was Karu's"—a deceased woman's—"biological mother?" I asked. Three different women were mentioned. (It turns out that people could have several adoptive mothers, as well as a biological one.) One was finally agreed on, about whom I asked, "Who were her children?" I got a list of names, then was told, "em tasol" (Pidgin for, "them, that's all"). I noticed that the deceased woman was not included in the list.

"Wait a minute," I said. "What about Karu?" More discussion ensued. Sayuma then nodded, smiling as if everything were fine. "And Karu," he said, translating. Everyone smiled, looking innocent, helpful, and nonchalant, as if nothing had happened. Fortunately these problems were the exception rather than the rule.

I was glad for a day of *malolo*, Pidgin English for "rest."

In the morning, Sayuma came by and asked if I wanted to buy his wife's kina shell for 5 kina. The simple shells—a dime a dozen in any beach town—were prized here in the Highlands, where traditionally they had been rare, traded all the way up from the coast, strung together, and worn dangling around the neck on a string as if diamonds or rubies. But they remained merely cheap shells to me. Sayuma lowered the price to 2 kina and I said I'd think about it.

I plotted out Fore family trees from my notes, but realized I was missing sections, usually because the informants didn't know the information. I realized I could take down the data more efficiently in the field on large taped-together sheets of paper, mapping the information out as a family tree. Otherwise, I had to recopy it as a family tree later. The data would also thus be clearer and more legible in the field. I saw now that I would have to go back to Purosa and Kanigatasa to get more detailed information. As I did not know if I was overtaxing my informants' memories or not, I tried making a family tree of my own to check. I counted sixty-three relatives whom I had met and known. I had met but didn't remember five others, mostly my grandmother's siblings. I realized I didn't know much about either my father's parents or grandparents—or my maternal grandfather.

I also classified all of the family histories I'd collected thus far into one of three categories—connections (i.e., to past kuru feasts), clusters (of more than one recent patient traced to a past feast), and incompletes.

Incompletes were patients for whom either I had found no additional close relative who also had kuru, or much information was still missing, since informants did not know parts of the family.

I had brought with me only two books—James Joyce's *Ulysses*, which I had started to read, and Jack Kerouac's *On The Road*. A friend later wrote, asking why I was not reading something escapist instead. But James Joyce kept me sane. The mental effort during the day here was intense, the sensations bizarre. To sit down to read a book reassured me, refreshing me with the taste of intelligence and insight as I used to know it. I didn't need to vacate my mind with escapist books as much as I needed to remind myself that intelligence could exist amidst chaos, that uncertainty and ambiguity abounded in the West, too, along with an underlying order. I could always chat with Maryanne and Roger, but picking up a Great Book jetted me back to civilization more readily and quickly, reminding me in all the confusion around me of culture with a long tradition, and helping me appreciate my current environment more.

It was very easy to walk around each day in Waisa, to buy vegetables at the door and limit my contact with the Fore to seemingly straightforward daily interactions. To think, see, and keep my mind, ears, nose, and eyes awake took energy. I had to keep my bearings in this world, during walks in misty forests, and arguments with superstitious men who feared that I could harm patients simply by looking at them. I was fighting entropy, ignorance, amnesia, and a gulf of thousands of years. I needed something to keep my mind alive, and without films, television, telephones, music, or radio (other than short-wave), books were it. I now appreciated how Westerners had more time to read books in the past, before these other diversions evolved.

Here in the bush, my time varied between those hours when I was trying to accomplish research work, asking specific questions, and trying to elicit specific answers and information in interviews, and those other moments when I merged with the environment or the people and observed and reflected on other aspects of the culture, including the contexts of disease.

Busakara, when he saw I was interested in local culture, came by one day and talked to me at greater length. I had begun to make lists of words I didn't know in Pidgin. I was also growing interested in *tok ples*.

He listed categories of words for me—dozens of names for each different kind of cloud, taro, *kaukau* (a large white yam–like potato), and *kume* (a very strong-tasting lettuce). There were ten kinds of sugar cane, and twelve words for chopping wood—hacking, butchering, splintering, splicing. Until recently, the Fore had only stone axes. Busakara left and returned to sell me two stone axes—black, oblong, and smooth—the last ones around. Once, not long ago, they represented the culture's most advanced technology. I would now use them as paperweights.

Maryanne and Roger were bored when I asked Busakara to elaborate on various words. But to me, these terms revealed what was unique here. To try to make sense of the place and my experience here was important to me. I had tried to take photographs, for example of a man splitting *pitpit* to weave into walls, of two women weaving *bilums*, and of a child playing with mud in a tin can. I wanted to get candid shots, capturing people unselfconsciously, and scenes as they were in reality. But the Heisenberg uncertainty principle generally operated: my entrance affected whatever was going on. My observation altered what I was observing. When I told a woman walking down the road carrying a great sack on her head that I wanted to take her picture, she froze, postured herself, and dropped her hands stiffly to her sides. All her *wontoks* posed formally, too. Old men liked to get into shots, and stood in the way, smiling for the camera, eagerly as they never did otherwise. I tried first looking and focusing on another object a similar distance away, setting the light meter, and then slyly and rapidly turning, aiming at the desired subject and snapping the picture. Still it was difficult. Words reflected and let me capture what was around me more readily than photos or video. Filmed documentation, though commonly thought to be objective, isn't necessarily so.

I realized the benefits of trying to describe what I observed in writing. In certain ways, writing about my interactions here could capture Fore life and mentality better than, say, a quantitative study of diet or disease. Not that these other approaches weren't important, too. But I saw how science itself results from data, and is shaped by the limitations of the data. It proceeds in small steps, each of which can't apprehend phenomena themselves, but only aspects of them. We cannot study the brain, the body, or a culture itself, but only their parts and outward manifestations.

I realized, too, that my official role here as "medical researcher" was enabling me to enter hamlets, homes, and lives, and to observe people in ways I would never be able to do otherwise.

That evening by the fire, I reflected on all this. Here in the bush, without the distractions of TV, countless radio stations, newspapers, and telephone, I had a lot of time to think.

That night, I thought, too, about my future. I decided that I wanted to live an active life, engaged with the world. What I had seen and learned intellectually from Carleton was living life. Literary critics and academicians missed the point by being too analytic and petty, I thought. I wanted to be a primary observer—not a secondary one who reflected on what other people have thought. As Montaigne wrote, "More books are written about other books, than about anything else." I wanted to see the world and nature for myself, observing the human mind and human biology and culture directly, more than simply studying literature, for example, per se. I was interested in how people's perceptions of the world and themselves flowed and changed in the context of different cultures. Yet I lacked confidence in this more intuitive, less systematic kind of thinking, to do it alone. Medicine appealed to me as an activity based on direct experience and interaction.

I remembered that Gertrude Stein had gone to medical school—as her professor, William James, had suggested—as had Rabelais, Chekhov, Somerset Maugham, Arthur Conan Doyle, William Carlos Williams, and, though briefly, James Joyce. I wanted to be free intellectually like Malraux, Thoreau, and Lewis Thomas, none of whom were academics per se. I hoped I was making the right decision. In the meantime, I would continue on here despite the difficulties, sensing that this work was important, and also that I was on an adventure, not knowing exactly what I would find.

I went to sleep, then got up to go to the outhouse. A blaze of white light slashed down through thin veils of purple clouds on the horizon. I crawled back to bed and fumbled for my watch: 1:30 a.m. What mysterious, natural event could this phenomenon be, throwing white light down beneath purple clouds? My mind raced. I suddenly remembered that the French were then conducting nuclear tests in the South Seas. Or was it some comet, streaking unknown across the skies?

I assembled my camera, loaded it with film, mounted it on my tripod, and photographed this fantastic occurrence, whatever it was. The stars sparkled in full brilliant clusters across the sky. I saw the Milky Way from the Southern Hemisphere. The ball of light was by then vanishing in streaks and sharp points of light between eerily somber clouds. I thought I could make out the Southern Cross, which I had only read about in the literature and lore of the South Pacific. Perhaps the ball of white light had only been the moon—rarely seen in these usually overcast Highlands. But I doubted it. (I never did find out what it was.) I felt I was braving darkness and fear and felt confused. I wanted order, explanation. That night in my dreams Roger calmly reached up and pointed out the Southern Cross to me. I was entering a new world.

Turn Talks

My guides still kept scheduling times, but then coming late, confi-dently and unfazed. I told them, as we started to have further destinations, that I wanted to try leaving on time. Sayuma and Sana as-sured me that that wouldn't be a problem. But the next morning, Sayuma didn't show up at 7:30, the time we had agreed on. After an hour, I shouted out to him and then sent someone down, who returned and said Sayuma was just washing up. Twenty minutes later he still hadn't appeared, and I left. An hour later, Sayuma caught up with me on the road, claiming a man had been cross with him.

"Well, were you ready to leave at 7?"

"Yes."

I knew this was a lie. His lies fed on themselves; he had no sense of the larger truth of a situation. But since I couldn't change him, I preferred that he be unable to fool me rather than that he become a better liar.

"We need other carriers," he said.

"I don't think so."

We set off to see Saroma in the village of Kalu. Sayuma did not of-fer to help me carry either my knapsack—which was filled with food and drink for the two of us—or a computer printout that Carleton had sent me from the NIH. Sayuma merely walked behind me as I trudged up the first hill. He carried nothing. Then I asked him to help. At first he pretended he didn't understand, then he said he would carry one of the two objects when we got on the trail. But he turned and kept going.

"What about until then?" I called. He kept walking. Selfish bas-tard! I realized he wanted me to hire and pay someone else as a porter to carry his bags, too. "Sayuma," I yelled out. "This is no good." When he saw I was angry, he backed down and took the box. I told him I was going to report to Carleton on my guides' cooperativeness.

Saroma, whom we had come to see, had been in Kanigatasa when I visited there, but the men assembled at the time had thought it best to ask her husband first before I saw her. Sayuma, trying to appear diligent, had made a point of telling me several times that he was going to come here to Kalu to ask permission while I was away in Goroka. But now her husband, as I talked to him, wanted me to come back to see her another time.

"Why?"

"She's in Takai," Sayuma quickly said.

I was suspicious. "Well, *can* I see her? I asked, wanting to get at least his permission.

"She's in Takai," Sayuma said again quickly. "You need to come back."

"Well then, I will go to Takai today to see her." I sensed something was awry, and liked being a stickler rather than tricked.

Sayuma said her husband would have to go and get her tomorrow, and that I should come back Monday. "Besides," Sayuma said, "you will be too tired to go to Takai." I knew for sure then that he was lying.

"I'm going there now!" I announced, standing up. I was still uncertain if I was pushing too much. Suddenly the husband called out to a hut. A little boy ran out, and a woman emerged from where she had been hiding behind a fence.

"This," Sayuma said, "is Saroma."

"I know," I said.

She walked a little bit, without difficulty, and sat down.

"She has malaria," Sayuma said.

"Malaria and kuru?" I asked.

Sayuma spoke to the husband, and then translated. "She does not have kuru anymore," Sayuma said.

This time I understood the husband well enough to see that the translation was correct. Saroma was stubborn, but eventually complied with my examination. Her husband was right—she did not have kuru. But why had Sayuma said she had kuru, and why had he claimed that he'd gone to get permission for me to see her? They said she had been cured. I assumed she hadn't been. But if not, how did they manage to believe this?

We went on to the next hamlet, which was close by, to see another patient. A short man saw us approaching and quickly skirted out into the surrounding bush and vanished. Sayuma called after the man, whom he said was the patient's husband. When we got to the hamlet it was deserted. Sayuma said we should wait. I felt we should go visit another patient in another, nearby hamlet. I thought we could then make a surprise visit back here later. Slowly, kids began to appear. Sayuma sent one down to announce that we were here. Fifteen minutes later, a man trudged up without looking at us. It was the man who had run away when we had first approached. Apparently his wife was inside a hut. He looked frustrated and depressed. I tried talking to him, with Sayuma translating. The man then got up and ducked inside a hut, bolting the door from the inside. Sayuma tried talking to him in *tok ples*. Through these bamboo walls—more like curtains or screens than the walls of houses—every noise and voice can easily be heard. The man sulked, but ten minutes later came back out. I could look at her, he said, but could not take pictures. What superstitious people, I thought, believing that photographs could endanger health. Ignorance breeds fear and vice versa.

The woman stepped out of the hut. She stumbled, and I immediately observed an ataxic, or uncoordinated, gait. I gathered an extensive family history. An elderly man exchanged friendly looks with me, then joined us and helped. I learned that the patient's brother had died of kuru two years earlier, and that two of her stepmothers had died of kuru right before White men came. The stepmothers almost certainly would have been consumed, probably by their stepchildren. Unfortunately, the villagers didn't know anything about the stepmother's families. But, I was told, I could probably get information in Mentilasa, where the family comes from.

While in the village, I decided to look for the husbands of two kuru patients who had recently died. The elderly informant advised me to go to a nearby hamlet where Roger was working. I was a little confused, but it was on the way. When we got there, the whole hamlet was gathered and the women were cooking. Sana told me, "This is a large *mumu*"—a feast. Men were working on the washhouse Roger was installing. Women prepared food, wrapping dozens of bundles of vegetables and freshly killed pigs in broad, velvety green banana leaves for the *mumu*. In the past, human flesh would have been added, too. It started to rain a bit. The men moved the huge pile of bundles onto rocks and coals left over from a fire at the bottom of a pit. They then dumped dirt on top of the bundles, filling the pit and creating a huge mound. Steam filtered up through the loose soil. I realized that once it would have smelled of cooking human flesh.

Sayuma had sent someone to fetch one of the husbands for me to interview. After an hour, the man still hadn't come. His son then called out across the valley in a loud, reverberating voice. In the far distance, a yodel rose up from somewhere in the trees in answer. Finally, he arrived, glad to meet me and to be the center of attention. The tribespeople contacted neighboring villages through an amazing calling system, projecting their voices and maintaining extensive lines for sending messages by mouth from one person to the next.

I conducted the interview in the half-completed washhouse. The interview went well, though the surrounding mud, and the rain falling on my back, head, shoulders, and notebook distracted me and made me uncomfortable. Soon, we were finished. A large crowd had assembled out of curiosity and to keep dry, but many had lost interest once it was clear I was just asking basic questions, with little novelty or unusualness. I, too, was glad when the interview was over. I shook everybody's hand. An old man wearing only traditional dress—a few leaves around his waist—hobbled up to me and nodded. He mumbled something in *tok ples*. I didn't understand. "He said," Sana translated from *tok ples* to Pidgin, "Do you know, 'New York City—she's a friend of mine?'" I looked confused. "It's a song on the radio, on the short wave," Sana explained. A villager had gone to work on a plantation on the coast and had bought a radio which he had brought back with him when he had returned briefly to visit the village.

Another man asked if we would like to join the *mumu*. I asked Sana if he thought it would be okay, and he said it would. I was glad. Another man went to get Roger, who had gone on to another hamlet. These people were very generous, and more open than those at Waisa.

Twenty minutes later, I was surprised to see Roger come up the hill toward us—someone I knew, another Westerner, here in the middle of the wilderness. He seemed almost out of place.

Making a mumu.

The men removed the mound of dirt from the pit, using their hands and a shovel. Traditionally, they were unable to boil water, as they had no pots or pans. The men pulled up the steaming bundles of food, which the women unwrapped. The food got wet in the downpour, but was delicious anyway. The steamed *kaukau* was piping hot and soft and especially good in the cold rain. When peeled, an edible variety of *pitpit* tasted surprisingly like asparagus. The beans were good, but were still beans, the pods dirty. The outside of the *kaukau* was also unwashed and covered with earth. I cautiously put the skin on the ground, so as not to attract any attention. A dog saw it and skulked over. The whole village sat quietly and ate, but conversation was absent. Nor was there much to eat compared to one of our big meals.

Afterward, people got up. Two adolescent boys, at an age that made them curious about life beyond their village, wandered up to me. "Em i bigpela kaikai" ("That's a big meal"), one of them said.

"Yes," I answered.

"After a big meal like that we go to bed right away," the other boy added.

It was around two or three in the afternoon. The conversation saddened me. This dinner was just one of my meals that day, but it was their only big meal in several days. Sayuma even said to Roger and me, "this way, you won't have to eat when you get home." I felt sad at the limitations of their world. New Guineans weren't fat. But when Roger and I got home, I cooked dinner. I made potatoes and fish cakes—potato latkes with canned fish added into the batter, and some fresh peas for what I thought of as "ethnic interest," to put some texture and variety in the otherwise homogeneous mush. I also cooked fresh carrots, tasty with margarine melted over the top.

After dinner, Sayuma came by. "Tomorrow, we will need other carriers," he said. I looked at him now, a Band-Aid above his eye, his gums decayed and eaten away, exposing a handful of rotted, yellowed and brown teeth with their roots fully visible. His nose was pierced to allow a small stick to pass through from one nostril into the other. His ears were also each pierced twice. Ugly, bony bulges protruded from his temples and his cheek, which caved in below his puffy, fleshy face. His eyes poked out from the corner of holes in his flesh. He looked like a pirate, and indeed I sensed a certain deviousness in his eye.

"I don't know about that," I said.

"But the village we're going to is far."

"Okay," I finally said, to quiet him. "We can take one extra person."

But in the morning when we got ready to start off, he had brought four additional carriers—all of whom would want to get paid.

"Why are all these carriers here?" I asked.

"You said it was okay."

"No I didn't. I said only one."

"No, you said three."

"I did not. Besides, you have four here!"

"But you said it was okay."

Sayuma's pushiness increasingly irked me.

We took one and walked. When we arrived at the village, I was told I couldn't see the patient. I asked Sayuma twelve times why not. I got twelve irrelevant responses. All in answer to the same question.

We trekked to another, further village and saw a patient who, Sayuma told me, "has had kuru and been cured six times." I was dubious (in fact she turned out not to be sick). As we hiked back, Mr. Lewis drove by and offered us a lift. I gladly accepted. But a few miles down the road we came to a landslide that we had to shovel out to enable Mr. Lewis' truck to pass, moving pounds of dirt, straining my arms. Afterward I rode on the running boards for a while, supporting my whole weight by reaching inside the driver's window and grasping onto the sharp metal edge of the truck's roof as we rumbled down the bumpy road.

In the morning, my arms were still aching as we hiked to another patient's hamlet. Her husband looked frightened and shy. The interview was too much for him—he refused to allow even Sana to look at his wife, Nauye. I didn't know if she was actually ill, but suspected she was. I told myself I'd keep coming back until he acceded.

The next patient that day was a woman whose brother let Sana and Sayuma look at her, but refused to let me. "But I am trying to 'win' kuru, to find a cure." I said. The patient's brother still looked tense. "Kuru research has been going on for fifteen years, but that is a short time in the West," I said. "This truck, this watch, these clothes, airplanes, take upwards of fifty years to invent. What would have happened if people in America had taken the attitude which you are now taking, and refused to allow people, doctors, to try to work on these things after fifteen years? There would be no planes, trucks, roofs, or anything. If you refuse me now, over the years more people will die of diseases like kuru, here and elsewhere." (Admittedly, a slight exaggeration, but the principle was, I felt, right.)

"Do you want other children, and their children to die of these diseases?" I continued. "If so, you are right to refuse me. If, on the other hand, you do not want to see them die, you should let me see your brother. Secondly, my looking at the patient will not at all harm the patient. I am looking at you, and others have looked at the patient."

Finally, I won them over. "Bel bilong mi em i tait" (from the English, "My stomach is tight," as *bel*, meaning belly or stomach, is considered the seat of emotions and mind; the phrase thus means, "I'm feeling tense"). "But you can go back to see the patient," the brother said. Sayuma and Sana proceeded to be of only minimal assistance, especially with the first two family members. Sayuma, in particular, seemed almost to want to encourage the resistance of families. In retrospect, I felt badly about my assertiveness, but that their resistance was based simply on ignorance. I didn't think my seeing the patient would harm her. The woman had kuru, but I couldn't get much information about her family.

A few days later, Sana came by with a transistor radio. "Em i baga-rap," he told me. (From the English, "bugger up," meaning it had broken). "Here," he said, handing it to me. "Fix it." He assumed I'd know how to. I opened up the back. A transistor battery was snapped in. I replaced the battery with a spare that Roger happened to have in the house. But the radio—an inexpensive plastic one—still didn't work. I handed it back to Sana. "I can't fix it," I told him.

He shook his head as if I didn't understand. "No," he said. "Fix it." He couldn't comprehend how I wouldn't be able to. After all, he knew everything there was to know about the technology in *his* culture. How could I not know everything about the technology in *mine?* He wouldn't accept the radio back until I had repaired it. "I'm sorry," I said. "But I can't." He looked at me as if I were either crazy or stupid, and reluctantly took it and walked away.

To travel further, I decided to hire a local PMV for a day. I told the driver, Tony, that I wanted to get an early start, and he agreed to come by at 7:00. I wasn't sure how he would know when it was 7:00, but he confidently said it wouldn't be a problem. I assumed he knew the sun rose at around 6:00. But he arrived at 11:00, with a truckload of twelve people. "I left my house at 8," he explained, "but the road was bad." I didn't believe him, and even if so, he should have left earlier to arrive at Waisa by 7:00.

We finally got on the road to Kasoru. A patient there, Pei, had recently died. I took down a detailed family history, though I was unable to find any other recently deceased victims in the family. Tony then wanted to drive on to Ilapo to take a woman there. I said OK, and decided I would try to get histories of two patients while we were there. But unfortunately their *wontoks* turned out to be away.

Finally we returned to Waisa. Sana then announced that he was going to take his water tank in the truck from Waisa to Purosa. I calculated the cost of us riding from Waisa to Okapa, to Kasoru-Okapa and back, at fifty toia each way. Tony had wanted to go to Ilapo, not I. I said it came out to six kina, but that I'd give him some extra money, up to nine. Tony was disappointed. "Carrying cargo gets thirty to forty kina a day!" He had gone only fifty kilometers, according to the odometer.

"Just fill out an Institute form," Sana told me.

"But I'm paying myself," I said. I resented being a tool for their economic benefit. I thought I would deduct carrying the twelve *wontoks* and Sana's tank, and the woman to Ilapo. They saw me as a font of money as if magically, reflecting cargo cultism. New Guineans believed that White men's cargo or goods were sent by ancestral spirits, intended for the natives, but that Westerners, knowing secret incantations, had intercepted the goods. Over recent decades, cargo cults had sprung up throughout New Guinea. Natives had built their own imitation airstrips in the jungle, thinking planes would now land and bring them goods, too. On the coast, a Lyndon B. Johnson cult had formed during his presidency, after hearing that he was the most powerful man in the world. In New Hanover, the natives elected him chief. When he didn't arrive, money was collected to offer him. Recently, many Fore men had built *stuas*—Pidgin for "stores." These men had seen Westerners build such *stuas*—with locks placed on the doors—that were soon stocked with goods. Native men now erected buildings themselves and put locks on the doors, and assumed these *stuas*, too, would soon fill with goods and make the men rich. When it didn't happen, the natives assumed not that their underlying premise was wrong, but that they merely did not yet know the secret—the incantation—that White men and women possessed.

New Guineans thus believed that Western goods they wanted were rightfully theirs. The Fore did not expect that they needed to deal with me in a reasonable way, which I found very frustrating and disturbing.

But I didn't want to cause too much friction. In the end I gave Tony twenty kina. I told Sana and Sayuma that I wanted to walk tomorrow. They said they'd come by in the morning, and they headed off with the truck.

In the morning, they didn't show up. I suspected Sana wanted to install his tank. They arrived the following day, and pushed hard for me to rent a PMV again. I refused, but they pushed further. We heard a rumor on the road that Michael was coming to Waisa at two o'clock. I used this as an excuse, and they finally let up.

I went to a village to get information from an old man named Papoka about a recently deceased patient. Papoka wasn't around, however. I told his son, Alo, that I would come back. But Sayuma wanted to stay, hang around and talk to some of the men in *tok ples*, and wouldn't leave. I tried interviewing another old man about the kuru patient's family, but the man didn't know much. I ended up waiting around for Sayuma, which took over an hour. I realized, too late, that I should have used my better judgment, and come back later, or visited Miss McGill, who lived not far away. But I was able to dissolve into the hamlet's slow pace. I heard a story about a man who lived in the bush by himself. He was not a "man of clearings," and two snakes came, one of which the man killed. The other escaped. The man eventually died. That was the whole story. The moral: living out in the bush by oneself has dangers. While I waited, I asked Alo if he would want to be a bush man. "No," he said. "I don't know how to kill the animals." I realized that what held these hamlets and this society together was, in part, the danger of the bush all around. New Guineans can be fiercely independent. Surrounded by vast uncharted, unoccupied land, people are not forced to live together, but can easily go off and form their own village or hamlet (hence the wide diversity of languages). Yet even here, they go forth not as individuals, but as groups. The forest threatens them, yet keeps them together.

After a few hours, Papoka still hadn't arrived, and I left.

Sayuma said he'd come by and take us to visit a patient named Matema in Yagamati. But Sayuma didn't show up for several days. Finally, he arrived one morning—a day when Sana had said he would come as well. But Sana now sent word that he had to do something in Purosa, and couldn't make it. Sayuma brought with him a man slightly older than himself. "He is from Yagamati and knows about Matema's line," Sayuma told me, proud to have done me a service.

I sat down with the man, and Sayuma walked around to the back of the house. It turned out that the man didn't know anything about Matema. "You are from Yagamati?" I asked him.

The man looked confused. "No," he said.

"Where are you from?"

"Waisa."

I got up and went looking for Sayuma. I circled around the house and the hamlet. When I returned, the man had gone. I went inside and found Sayuma seated, eating. "That man's not from Yagamati," I said.

"Yes he is."

"He said he isn't."

"He is."

"Where did he go, anyway?" I asked him.

Sayuma paused. "Back to Yagamati," he said. Another lie, I could tell. I looked outside. The man was there again.

"Let's go to Yagamati," I told Sayuma. He didn't want to, but I insisted.

Later that day, I heard that Sana was seen going to Goroka.

"I won't come by tomorrow morning," Sayuma told Roger and me that night. "I'll go look for kuru tomorrow in Kania."

But the next day at noon, Sayuma nonchalantly strolled by the house.

"I thought you were going to Kania today," Roger said, annoyed. Sayuma came in and made something up now about his wives. I escorted him out.

I tired of all these problems. I wanted a day by myself when I could just sit, read, write, and reflect. I also had stomach cramps from something I had eaten or drank. Here in New Guinea I often had intestinal parasites from the food or water. Though I tried to be careful, even the tap water in Goroka was often found to be unsafe. Sometimes, I would have diarrhea for weeks on end.

In the morning, we heard that a funeral was being held for a woman in Ilesa. A contingent from Waisa—Sayuma, Busakara, Soba, and one or two others—planned to journey for the three- to four-hour walk, carrying a coffin constructed by Jake Lewis. I was tempted to go, never having seen a funeral here. Intellectually, I would have found the event worthwhile. But the afternoon breeze was cool and relaxed through the window of the house. I was glad for a break. If I went, I would have to stay in a hut for one or two nights and eat only their diet of vegetables and pig. I decided to stay here and rest. I turned on Roger's short-wave radio. It picked up only two stations. I listened for a few minutes to the Voice of America—filled with pro-American sentiment, virtual propaganda. It was my only contact with the rest of the world in weeks. Still, after fifteen minutes I was glad to turn it off.

A few days later, I decided to see another female patient near Waisa, and asked Sana to come along, since Sayuma was vague about her, and hesitant to see her. I was afraid Sana would be annoyed at having to travel all the way from Purosa just to spend one day in Waisa. But he seemed satisfied to come, because, he said, he would go to a *mumu* being held that day to celebrate a man's burial.

At this funeral, a man got up and made a speech. His family wanted payment from the deceased's family because of a woman married to the deceased. The bride's price had apparently never fully been paid. Intense discussion followed, and finally agreement. A pig was brought out, cut into three or four pieces, and given to the woman's family. Sana told me the man could have gotten gifts instead by having her "sold" (i.e., married) now again.

Afterward, a short woman wearing only a grass skirt hobbled up to me, looked me up and down, and nodded. I smiled back. Then she reached up and rubbed my arm, just above my elbow, mumbling something in *tok ples*.

"Em i toktok wanem?" I asked Sana—Pidgin for, "What did she say?"

"Em i toktok long arm bilong yu em i be gutpela long kaikai," meaning, "She said that your arm would probably taste very good."

I pulled back in horror. She grinned at me. Sana did as well. But I was wary. After all, she had eaten her share of human arms.

I smiled as politely as I could, and walked away quickly and uneasily.

A week later, a man from Paigatasa stopped by to tell me that Ganara would be coming soon to work with me as well. I mentioned it to Sayuma, who was surprised and then got nervous.

"We don't need him," Sayuma insisted. Sayuma acted very proudly, seeing himself as a hard-working man, though I knew he didn't look for kuru as much as he claimed. I just had to remember that it wasn't me he was ripping off, that I was losing nothing, and that my time here was limited.

The Last Thing You Can See

The following Monday, at the crack of dawn, Sana, Sayuma, and I hiked off on my first overnight trip. We were traveling to one of the furthest edges of the Highlands—the village of Agakamatasa, beyond which uncharted land stretched for hundreds of miles.

The night before, I had packed the essentials—bottles of the hottest spices I could find ("extra hot" curry powder and Mexican-style chili powder), Band-Aids, a camera, pen, and paper.

The first night, we stayed in Purosa with Sana. He had 17 children—13 of them adopted. He believed they should all attend school—a plan that seemed inspired by and in homage to Carleton. My respect for Sana soared as I viewed his family and his life more closely. Staying in a man's house enables one to see into his heart, and his worries. Sana's adopted adolescent boy, Jim, tried hard; he was on the margin of being a fully accepted family member. Sana's wife, with a soft, tender, slender feminine face, spoke perfect Pidgin—unusual for a woman here—though she was shy, as was the custom of Fore women. Sana was a "big man" in the village—a leading citizen in this until-recently Stone Age town. As a gift of welcome, he gave me four arrows he'd made himself. Sana took very good care of me, setting up a hurricane lamp for me to use at a table, and a blanket and pillow to sit on for a chair. He then brought me a delicious and satisfying dinner of three juicy cobs of corn and a heaping plate of rice and fish, to which I added spices. The spices made me thirsty. Luckily, I had a full flask of clean water.

In the morning, we hiked and after several hours emerged on a mountain top clearing. Far below us lay a village—Agakamatasa, the last set of hamlets at the edge of the group's territory. To the south, as far as I could see, stretched untouched mountains and valleys.

"What's that way?" I asked in Pidgin.

"Nothing. No one lives there," said Sana.

"Why not?"

"There's no reason to. Everyone already has enough land here—more than he can garden."

Weeks later I happened to see a topographical map of the region. I noticed that almost all villages lay in valleys, and that no valleys existed south of Agakamatasa—only mountains.

We passed a collection of buildings known as a *singsing*—forming an enclosure where celebrations were held, for example after the rainy season. In a fashion recently sweeping the highlands, villages had built *singsings* for feasts with beer drinking and singing. Yet these events had become rowdy, and missionaries and the government had tried to stop them. In any case, this *singsing* consisted of three buildings, each looking very different: one was crudely constructed, another neatly. One had a low roof, another a high one. One was tall and thin, another short and fat. "Why are they different?" I asked Sana.

"That's just the way they are," he replied.

"But it's odd," I said. "There's nothing to explain it?"

"No."

"Were they made by three different men?"

"Yes," he said. Yet it did not occur to him to say this at first. To him, they were all the same—*singsing* houses. The Fore didn't look at them the same way I did.

We continued to stride downhill as a surprisingly fresh breeze swept by, rejuvenating me. The village hung at the edge of a cliff high above the Lamari River valley. The river drained the New Guinea Highlands—one of the world's densest mountain ranges. The peaks solemnly gathered together, shoulder to shoulder. Directly across the river, an almost vertical wall of mountain faced us, the steeply chiseled side carpeted with trees and grass. I spotted a few huts—tiny in the distance—and a single plume of smoke. To the south, along the course of the river, lay a spectacular view of turquoise hills, ever paler and more distant, cascading one after the other from great heights, down into the gorge. At the very end, in the far distance, hundreds of miles away, wedged between the last of these tumbling mountains, lay a shining, pinkish-blue triangle—the top line perfectly horizontal. That, I realized, was the ocean—the Coral Sea—separating New Guinea from Australia. The cool breeze I had felt was in fact a sea breeze. I filled my lungs, amazed to be smelling the fresh saltwater air even here.

"That's the ocean," I said to Sana excitedly, pointing.

He didn't know what I was talking about.

"There, the last line—the last thing you can see—it's a big body of water." Sana looked down the river valley, perplexed. Then he glanced across the river at the mountain immediately facing us, sloping up from the river—land belonging to his tribe's traditional enemy, the Kukukuku, the fierce band of warriors who traditionally didn't even have a name for themselves. Sana looked back up at me. He couldn't figure out why I cared about some distant line barely discernible—almost a cloud or mirage—while across the river in clearly visible huts lived his enemy. I imagined him thinking that these White men made no sense. I no doubt confused him as much as he did me.

As the afternoon passed, dense, fluffy white clouds—different from those usually seen here in the Highlands—began to roll up from the sea in a long procession, rising up from an invisible horizontal line. I wondered all the more at the Fore in this region, looking at that triangle of light and this long line of cloud marching overhead each day for millennia. Yet the tribesmen never knew that they were seeing an ocean, or that such a thing as an ocean existed, or that they lived on an island, or that the world consisted of anything other than the valleys they and their neighbors inhabited. They didn't travel on the Lamari River—it was infested with malaria-ridden mosquitoes, anyway—but they never seemed to wonder where it went or what that final shimmering field of blue was. New worlds can exist under people's noses, and go unnoticed. In the same way, Westerners, throughout the Middle Ages, had thought that the world was flat, and had missed scientific discoveries as a result. I wondered how many things we in the West still looked at and didn't see for what they were.

In this village, Carleton had built a hut—a base where he could store supplies that he had brought out over the years—books, paper, pens, kerosene, and tins of food that would last for years, and that were otherwise unavailable. He had visited every year in the past, but now did so only once every few years. This house was his fourth in Agakamatasa—all structures wash away here after a few rainy seasons. He later built another hut on the mountain top clearing from which I had first peeked at the village, and which afforded the most spectacular views around. He had once told me that he wanted to retire to this spot, unreachable by phone or fax, to write, read, and think without interruption. I visited Carleton's present hut, and found several stored boxes of books—classics by Dickens, Conrad, Nabokov, and medieval Chinese poetry by Wang Wei. I picked up the last of these and sat reading as the sun slowly sunk over the long valley into the sea. At one point Sana strolled over and I put the volume down. He looked perplexed by what I was doing.

"This is Carleton's," I said.

"From his house here?"

"Yes."

Sana picked the book up and looked at the opened pages. He gazed for several minutes at the endless mysterious white sheets covered with tiny black markings. He wanted to know, but couldn't comprehend what occupied Carleton's and now my thoughts in a little black and white chunk of woodlike substance in our hands. He turned several pages, peering at each of them for several minutes. I sensed that he realized the work Carleton and I did here constituted only part of our concerns, knowledge, and time.

In the early evening, rain began to fall. Koiya, who watched Carleton's house, brought us each a meal of corn, *kaukau*, beans, and edible green, leafy *pitpit* stalks. To the bland and pulplike *kaukau* I added a little seasoning at first, and then more with each bite—creating an odd and exotic blend that added some flavor, but also made me thirsty. eventually used up both bottles of spice on one potato, but it was now edible.

As we sat eating with the cold rain falling outside, Koiya told me that as a reward for his work, Carleton had taken him to visit America

"What did you think of it?" I asked.

"Nambawan country. New York nambawan ples" (place). Sana and Sayuma listened eagerly. Koiya turned to them. "Plenti people na plenti tall buildings" ("Plenty of people and plenty of tall buildings") They nodded their heads. But then he paused, confused. Koiya could not begin to convey in Pidgin or *tok ples* what he meant. He lacked words to describe what he had seen. He was relieved to talk to me, since I knew about it.

"What did you like best?" I asked, trying to help him out.

"Louisiana. It has bush," he explained—woods in a warm climate, with which he felt at home. Orange *kaukau*—sweet potato—had also amazed him. He showed us photos—mostly of parking lots, of which he had taken many pictures. He had never seen as many cars in one place before. Sana glanced at the photos, became anxious and overwhelmed, and escaped into the woods. I felt badly for him.

"What was it like coming back?" I asked Koiya.

"Hard," he said. "Everyone wanted to know if I had learned the secret of the White man's cargo." When he returned here, Koiya had been questioned hard as to the special prayer or procedure White people must have for getting their goods.

Koiya also showed me a stone sculpture of a cassowary bird he had dug up in his garden. The natives now revered the smooth, large white carving made from a stone that doesn't exist here in the Highlands. It had been made by a previous culture in this region. I would soon see as well unearthed fluted mortars and pestles. This earlier culture, long since gone, apparently possessed technology that these present inhabitants lacked. Who were these lost peoples? When did they live? What else did they make or do? No one knows the answers. Only these stone sculptures remained from this society. The Stone Age had come here after a more sophisticated culture had died out. This relic seemed like something right out of *Planet of the Apes*.

I had assumed that over time, cultures invariably advanced and the human condition steadily improved. I now saw how they didn't always. When one culture dies, a less advanced one may flourish in its place. Ancient Rome had, after all, succumbed to barbarians. Progress is itself a culturally created notion. Cultures depend on this belief in their own advancement, in their own superiority to prior cultures—otherwise why put up with all the inconveniences? Yet lost civilizations, lost knowledge, hold an allure. We, too, will one day disappear.

The next morning, the sunrise was spectacularly beautiful out toward the ocean. I could feel the water's presence beyond the mountains. Eventually it would take me away to more familiar scenes— lands of my dreams. Still, there was beauty midst these Stone Age basics. I was indeed in a new wonderland. No matter that none of my heroes had been here, or that nineteenth-century explorers placed Africa more on the map of my imagination.

That morning, we set off to see a patient in another hamlet. Here, the women and children were even shyer than elsewhere—having seen almost no White people before. Women avoided looking me in the eye and walked with their arms crossed before them over their naked breasts, their hands clasping their shoulders.

I entered the hut of the patient, a man named Owa. He lay still except for his hand trembling slightly at his side. Otherwise he was unable to move, and seemed close to death.

"Owa?" I asked. He didn't respond. I examined him as much as I could. Neither he nor any of the other cases I had seen displayed the emotional lability or "laughing" that had prompted the term "laughing death." I wasn't sure why this symptom wasn't apparent. Was the clinical picture changing? It wasn't clear why it would.

I stepped back outside. A light drizzle had begun to fall. I sat down with the older villagers. The oldest man in our circle wore nothing but blades of grass around his stomach. I gathered a family history.

As I sat there interviewing, the rain increased and soon soaked my head, shoulders, boots and notebook. But there was nowhere indoors to go in the whole hamlet—only dark huts. Also, I wanted to record as many details as I could.

A few days later, back in Waisa, Ganara arrived for the first time. I had been here now for eight weeks. He shuffled a bit on the front porch, and I invited him in. Michael had told me about him, but I had been busy with Sana and Sayuma and hadn't thought about him.

I gave him a tour of the house. "This," Roger explained in the kitchen as he lifted the tight-fitting lid off a large metal garbage can filled with flour, "is our *kaukau.*" Indeed it was. Ganara, however, didn't really understand. "It's what our culture is based on," Roger added. Ganara still seemed oblivious, which Roger didn't seem to grasp.

"I want you to come and visit my village and stay a long time," Ganara said to me, "and build an Institute House there." We arranged that he would come by next Monday and that I would hike with him toward his area to see several patients. "And we want an Institute House," he repeated.

"Building an Institute House is up to Michael, not me," I said.

"But the people of Paigatasa want an Institute House."

"I understand, but it is not my decision."

"An Institute House must be built there!"

"You'll have to talk to Michael about that."

Ganara returned late Monday afternoon. I arranged to go to Paigatasa the following day. To save Ganara the trip there and back in the meantime, Roger let him stay over in our house. Ganara was wearing old, dirty shorts with stuffed pockets. "You'll probably want something to sleep in tonight," Roger said to him. "Here, you can use these for the night," Roger said, handing him a pair of gym shorts.

That night, when I got up to go to the bathroom, I passed Ganara sleeping on a blanket on the floor, and noticed that he had kept on his own shorts, and had not put on Roger's. In the morning, we left and set out for Paigatasa. Only later would I find out that Ganara had never returned Roger's shorts and had merely taken them with him.

Jake Lewis agreed to take at least two and maybe all five members of my patrol to Awande, which was part way. But he was leaving promptly at 7 a.m. I told my guides to come at 6:30, but at 6:45, none had showed up. I stomped over to Busakara' house where Sana was staying. Rain was pouring down. "What are you waiting for? Let's go."

"I'm waiting for Sayuma."

"But we have to go."

Jake Lewis was leaving. I told Sana to stay and get Sayuma. He sauntered away, downcast. At the very last possible moment, he ran back, saying Sayuma was coming. "Hurry up, we're leaving," Mr. Lewis shouted.

As Mr. Lewis started the truck, Sayuma ran up and leapt onto the moving vehicle.

In Awande, at the turn off to Paigatasa, Mr. Lewis stopped the car. I jumped off with my guides. The other passengers also piled out. We all began to lift our bags out of the back.

"Isn't that my bag?" I asked Sayuma, pointing to a black satchel one of the women took as she climbed back aboard for the rest of the ride.

"No," Sayuma said, shaking his head authoritatively. Mr. Lewis quickly drove away with the other passengers. We were left gathering our things in the mud, and I realized then that it had been one of my bags. (Later when I tried to retrieve it, no one admitted to knowing anything about it. Someone obviously just kept it.) Not far away stood an old kuru hospital that Carleton had built years earlier to confine and follow patients, before the disease's mode of transmission and incurability had been discovered. I was curious to see the building, and also wanted to replace the lost items. I hiked to the hospital in the wet darkness, and called out. No one answered. I pushed open a window and climbed in, entering a room, closed off and deserted for years. Against a wall stood four metal washbins. Wooden platforms, once used as beds, now lay carelessly strewn and tumbled across the floor, looking like emptied coffins. Raindrops dripped onto the window ledges, echoing through the room. Carleton had tried other treatments here, but none had worked. The ghosts of countless patients who had died here seemed to linger. I replaced the silverware, cups, dishes, bowl, toilet paper, Band-Aids, salt, and seasoning.

My assistants had all taken too much personal gear; Sana and Ganara were each lugging two huge, overstuffed bags. I now needed another carrier and had to hire kids to help, which Sayuma organized. Seven came. But after fifty minutes of hiking on the road, they stopped and wanted to get paid and return home. I explained to Sayuma that in the future I would only take one or two, and that they would have to come all the way.

L ater that day, we passed through the territory of Oma-Kasoru. "You're going to Paigatasa?" villagers asked us. "Be careful. We are at war with Ivaki," a village en route. "They poisoned our water. Six women here have died." The whole village was in mourning, assembled in a large circle, finishing a *mumu*. "We plan to fight back."

We kept hiking. A few hours later we reached Amurei, which was allied with Ivaki—the other side of the feud. "A man from Oma-Kasoru stole an Amurei woman," a man there told me. "We fought back and won." They were celebrating the end of the battle. I felt I had crossed the Trojan wall from one enemy camp to the other. Passions completely reversed.

By mid-afternoon, the equatorial sun blazed fiercely overhead. I remembered that the vernal equinox occurred at this time of year—late March—when the earth swung closest to the sun, which then burned directly over the equator. I'd never felt the sun searing straight over my head as painfully before. In New York, even at its brightest, the yellow globe shines more softly from lower in the sky. Here, the ball of fire squeezed down fiercely, directly on top of my skull. My body created a small shadow pooled around my feet. The sun buzzed like a huge insect above me, slurping up all moisture from the earth, every slimmer of cloud in the sky. The heavens turned to a hard, limitless blue. My legs heaved slowly, heavily, one after the other, up and down hills. My boots lumbered forward, like unoiled machine parts, no longer attached to my feet. My right boot chafed and irritated my foot, which began to ache with every step. Everything grew silent except for my boots shuffling painfully along, the heels scuffing on the dried, dusty ground. Birds ceased their cadenzas, crickets stopped. Leaves no longer rustled, but hung limp, weighed down by the sun.

We had long since run out of water, and a sticky glue now coated my mouth. My throat felt as parched as the cracked ground. I couldn't squeeze any saliva into my mouth, and I began to crave water as never before. But there was none for miles around.

We trudged on. Finally, after several more hours, Sana turned into a niche along the road. I ducked inside and faced a clear crystal waterfall. In a shady cave of ferns and bushes, the fresh water sparkled, dashing down and washing over shining smooth rocks. The sight gratified me as few things ever have. I ran beneath the cascade, and stood as the icy torrent slapped against my skin, soaking my head, clothes, and body. At the base of the falls, the water swished, bubbling, trickling, and plopping all around. This orchestra of sounds echoed through my weary skull. I began to feel like a new human being, alive again.

As we neared Paigatasa, we passed a few people on the road, and they stopped to talk to us. They were surprised that we had made it safely through the Oma-Kasoru lands. "Are you visiting Joe?" they all asked me.

"Joe? Joe who?"

"You know—Joe. He's a White man, too," they said—as if that justified our knowing each other. I was told he lived in a village beyond Paigatasa. I didn't know him, and was surprised that they assumed I would.

We hiked for several more hours into the twilight, when rain again began to fall.

We pushed on as darkness gathered and night descended. Falteringly, I followed Ganara's white sneakers in the pitch black. No pictures or films of the Highlands taught me more about the region than these steep trails rising at seventy degree angles, and the sticky mud sucking up my boots. No wonder seven hundred and fifty languages were spoken here, and very little contact existed between groups.

At last, in exhaustion, we reached Paigatasa. The village had had only slight interaction with the West, and that had come only since the 1960s—wholly within my lifetime. Some children still had never seen a White person. The entire village owned only one coffee grinder, one battered metal pot, two blunt knives, and a handful of old clothes. Only one man wore a shirt—its front tails dirty and tied together. His curly hair protruded stiffly from both sides of his head beneath an old, dirty brown beret. (He and others, I later learned, used pig fat to mold their hairdos.) Men and women squatted on the dirt, taking turns delousing each other.

We stayed with Ganara. His house, a two-hour walk from the end of the road, hovered on top of a steep hill. The "road" itself had been impassible for several years. He had gotten a corrugated iron roof from Michael. Ganara had been told, when he built the house, to leave an overhang, creating eaves. But he had wanted to build as big a house as he could, and thus aligned the walls flush with the edges of the roof. The walls, thus unprotected from the rain, were rotting away. The house was now about to fall apart.

The house itself contained only walls and a candle. There were no windows or furniture whatsoever. Ganara was a short man, who made slow movements, as if time didn't matter. He served us only *kaukau*— no beans, corn, cucumbers, or water. We ate outside surrounded by eleven pigs munching grubs and grass from the muddy earth. The animals' greasy noses slid around me, as saliva-filled mouths sloppily snorted and licked. They'd probably slobbered over the very spot where I sat, I reflected, spewing out worms and germs. The natives here prized pigs as wealth, using them to buy brides, among other things. The animals—investments, rarely eaten—ran around freely. Half wild boar, their stiff black hair stuck up all over their bodies as they gruffly charged past huts, and in and out of woods.

The smells here, too, were further from the civilized world and closer to the Stone Age—including body odor, since people rarely washed and never used soap. The whiff of wood smoke hung over everything. Contrary to the romantic image described by Jean-Jacques Rousseau, mankind in a state of nature away from civilization does not lead an easy life. I felt disgusted and lightheaded.

It was here, too, that cannibalism had persisted the longest.

I was surprised to see that Ganara's family lived in the worst squalor of any I'd seen, though he collected from the Institute 30 kina a fortnight in pay—about $48—probably more than any other man in the area. Moreover, before White men arrived, the region had survived without any money at all, since gardens made everyone self-sufficient. Yet his kids walked around in rags. Pigs scuffled and chewed at his doorstep. I later found out he had recently refused to pay either for his wife to go to the hospital because it cost 1 kina, or for his son to go to school because it cost 5 kina. He apparently loses all his money in cards and beer. He and other men sit and play poker daily, expecting money to come not from hard work, but from winnings resulting from *maselai*, or spirits. The spirits will reward them or not, regardless. Meanwhile, his kids run around crying until his wife—a rough, unfriendly woman—screams at them coarsely. They then stiffen into silence, only to start bawling again a little while later.

I demanded water and finally felt a little better.

Should I be more accepting and forgiving of my assistants? I told myself that my frustrations resulted from the differences between our cultures, but it was still tough.

A man from the next hamlet came in, asking me for a washhouse. I explained I was not in a position to give one. He brought with him a *karuka*. This one was dried in the sun, rather than cooked in coals, and hence without raw ashes—which had dried and irritated my mouth—clinging to the seeds. Hungry, I fished out handfuls of the edible nut-like seeds.

Another man arrived. "The village wants the Institute to build a house here," he told me.

"Yes," Ganara said. "We want a White man to settle here.

"I'll speak to Michael Alpers about it," I answered.

"We want the Institute to build a house." Ganara repeated.

"But the Institute did buy you the galvanized iron roof," I pointed out. "In return for which you said you would let the Institute use this house when necessary."

"But I also want a stove, lanterns, kerosene, pots, pans, silverware, and blankets," Ganara told me—all of which he would presumably use for himself most of the time.

"You will have to speak to Michael about that," I told him.

"On Sunday," Ganara told me, "people will come asking for an Institute house." They already had. Presumably they wanted the runoff goods—money, tanks, and so on, that would trickle down.

Ganara also showed me the *stua* he had built, the lock still on the door.

I went inside his house. Sayuma's voice boomed outside where my guides sat. I had brought a Coleman lamp and set it up to write in my notebook. The entire hamlet turned out to watch. They had rarely seen someone writing sentences. They gazed at me intently, as if waiting to see some magic occur before their eyes as a result of this activity. As I crouched in Ganara's bare, roofed hut, their stares at everything I did disconcerted me. I now saw the wisdom of Carleton's approach to these situations—engrossing oneself in the task of writing, to the exclusion of other concerns.

That night I dreamt I was on Broadway and 74th Street, shopping in a grocery store and seeing boxes of frosted doughnuts—food I wouldn't be able to eat for months.

In the morning we got ready to visit patients. My foot, now inflamed, no longer fit into my boot. Instead, I wrapped the foot in a double plastic bag. I looked ridiculous, but had no choice.

"Make sure you arrange to buy some other kind of food for dinner for tonight," I told Ganara before I limped off. He turned and yelled something out into the valley.

"That's all you need to do?" I asked in surprise. He simply nodded.

I thought I'd first go to see a patient nearby, and then at the end of the day—or, if time didn't permit, tomorrow—I would visit another patient's family, who lived further away. We started hiking, though I lagged even further behind because of my sore foot. Three hours later, we arrived at the first patient's hut. When we got there, however, the patient's brother refused to let me meet the woman.

"I just want to see her quickly," I said. "I've come here from America to try to understand and help others with kuru."

"No."

"Can I just take a picture then?"

"No."

"Why not?"

"Pictures are dangerous to her health."

"Well, can I just get a family history?"

"No."

"Can you give me her parents' and brothers' and sisters' names?

"No."

"Okay, what is your name?"

He wouldn't tell me.

"Why not?"

The man just shook his head. He feared that writing down people's names would somehow harm their spirits. He said his wife had been poisoned, and that he was first going to have a native healer try to cure her.

I left disappointed, having trekked all this way. Native beliefs were frustrating me. As it was getting late, we trudged back, having failed to see the patient. My stomach now growled and tightened—straining on the water and *kaukau* on which I had had to survive. For the first time in my life, an image of a loaf of bread rose up in my mind, though I wouldn't be able to eat one for days.

We passed through a hamlet where the inhabitants were eating beans. I was tempted to ask for some, but assumed Ganara's family would have these for us, too. Finally, we returned to Ganara's house. My muscles ached, and my foot, dragged through miles of mud, was sore and tired. Hunger now tore through my stomach with warlike vengeance. To my dismay, Ganara had failed to get any food other than two *kaukau* for each of us. Not even beans. I had eaten half a *kaukau* this morning and nothing else. (Ganara, I found out, doesn't buy seeds for vegetables, and thus has only corn and *kaukau* in his garden.)

I had been reduced to living a Stone Age existence, and I felt like I was losing my mind. "You have to get something else!" I told him. He stalked off, and returned an hour later, having managed to find corn, cabbage, and sugar cane.

I was relieved. His wife began to cook it, but when I turned around she poured on half a leftover can of "dripping"—lumpy greyish-pink generic fat that the natives use. It looked disgusting, but was still food. She also poured on handfuls of commercial salt, as if to make up for the lack of any other seasoning—of which the Fore had none. (Traditionally, all they had was native salt made from reed or bark ash. The human body needs a certain amount of salt to survive. Yet this native substance was low in sodium, difficult to make, and not very plentiful.) When the meal was done cooking, however, neighbors came by and gathered around, all expecting some. I didn't know what Ganara had told them. But I got some. I put cooked, coated chunks of cane and cabbage in my mouth. They were greasy and salty, and I felt nauseated, but tried to eat a little more.

I went to bed hungry, and barely slept. I had more difficulty falling asleep than ever before, chewed and consumed by fleas and other insects, sweating under a sticky plastic sleeping bag, and feeling queasy, my intestines greased by the few spoonfuls I had eaten. I also feared catching lice, which infested everyone here. But I remembered Hazel's experiences and words. "This too shall pass, " I reminded myself. If she (in her 80s) could persevere and endure her ordeals in the bush, then I could survive mine. Finally, I fell asleep.

The next morning, I wanted to quit and go back to Waisa. But I had come here to gather data, and I'd been told a funeral was to be held that day in another nearby village where many elders from this isolated and more primitive area would be. I certainly didn't want to have to return here, and so decided to push myself now and go.

Finally, we reached the hamlet holding the funeral. The mourners all first covered their faces with mud, and then wiped it off to reveal themselves, to show they were not involved in the sorcery. A crowd of women wearing only grass skirts surrounded the coffin. The women lifted the box and carried it on their shoulders, strutting quickly away, emitting mournful sounds—not crying in tears and sobs, but wailing in disturbing tones like a Greek chorus. The men sat to the side by themselves, with stalks of grass or handmade pipes in their mouths. Had it been several years earlier, the body would now have been cooked and eaten.

Later, I interviewed a group of older women—known as *lapun maries*, meaning "old wives"—who together remembered a lost time, before the arrival of White men.

On the way back, we took a different route. We passed a rectangular house—the only one for miles around, as all the native huts were round. "Joe lives here," Ganara told me. I decided to knock and introduce myself, although I had never met the missionary. Joe opened the door. He was tall with light brown hair and a beard, which almost all male Westerners had—myself included now—since the scarcity of clean water made shaving difficult. He asked me in, and invited me to have dinner and stay over for the night in his spare bedroom.

"Are you sure that would be alright?" I asked.

"I'd be very grateful for the company," he said.

I gladly accepted. My guides smiled, and nodded amongst themselves. They seemed to feel that it was right for me to stay with Joe—that all White men knew each other and were *wontoks*. The Fore believed that not far beyond the biggest mountain on the horizon lay the rest of the world, consisting primarily of two villages—New York and Australia. These two places were near each other, they assumed, since almost all White people came from one or the other, and everyone knew everybody else—making these two villages just like theirs.

In the Fore social structure people trust and help only their network of relatives and friends—their *wontoks*. Unfortunately, this *wontok* system cut both ways. It was refreshing that people here distrusted formal bureaucracy. Such institutions were alien to traditions evolved over centuries in a culture closely tied to nature. At the same time, government corruption was rife, with politicians slyly finagling as much as they could for themselves and their lines. The leaders didn't have much of a vision of where the country should be headed. The country radiated with newness and an exciting lack of precedents, akin to that faced by Thomas Jefferson and the Founding Fathers in the United States—but in New Guinea, vision was in short supply.

My presence motivated Joe to prepare a turkey with stuffing and mashed potatoes, and to bake a chocolate cake—all using supplies he had stored in a deep freezer. These and other goods were air-dropped to him once every six to twelve months. He stored everything in the freezer, which was run by a generator—the only electricity for dozens of miles around—itself propelled by gasoline, also air-dropped.

I felt safe. I learned that no other Westerner had visited Joe in years. He was very kind, but religiously fundamentalist and a classic conservative—"though not as far right as the Birchers," he assured me. "I don't think there's a commie hiding behind every rock and that Dwight Eisenhower was a Communist. Still, America has become bad—evil. People get shot and killed there for no reason. Even here in PNG, law and order are not respected. Villages fight. After an Oma-Kasoru woman died of *tukabu* [poison] from the water, the villagers marked rats with the name

of each nearby village to determine who was in fact responsible." Joe
seemed to believe Oma-Kasoru's water had actually been poisoned. "The
villagers trapped the rats in bamboo cylinders and put them into the fire.
The animal who survived was said to represent the village responsible—
Amurei. The natives repeated the process, marking the rats with the
names of various clans, then the names of individuals. The guilty man
was later seen on the road and killed."

"We must teach them the truth," he said. "They still think in
terms of ancestral spirits. We tell our children to use reason. They tell
their children to do or not to do something because it will anger or appease the spirits." I thought Westerners implictly told children both,
but I didn't say anything.

That night I got to sleep in a bed under a soft quilt.

The next morning, my guides came by to fetch me to return home
to Waisa. They recommended taking a short-cut through the bush.

"It's a difficult trail," Joe told me.

"But it's much shorter," Sana said. Joe admitted it was. I wondered
how much worse than the road it could possibly be. I was eager to return home as quickly as possible and decided to brave it.

We were soon trekking—my foot still wrapped in plastic bags—up
and down endless mountains and valleys, through teeming, isolated
tropical rain forests. Thick green vines beaded with heavy droplets of
water draped over the trees. From the sides of tree trunks, new roots—
pink and dripping—slipped out and hung in the air, glistening toward
their destination, the fertile, soggy soil. Strange fungi, resembling red
potatoes with huge black eyes, clung to the sides of trees and shot out
tentacles. I felt like I had fallen into some strange lost valley.

My guides, having grown up hiking through such terrain, constantly ended up far ahead of me as I walked carefully over the slimy
roots, one at a time, trying not to slip into the calf-deep bogs. Still, I
kept losing my step, and fell several times. At other points I had to
walk through ankle-deep mud that sucked up and swallowed my feet.
With each step I had to pull my foot up out of the ground, making a
wet, sucking sound. Whenever I glanced at my calves, three-inch-long
leeches had crawled up and were stuck on my skin. Every few minutes
I had to stop and wipe off twelve or fifteen brown worms as their
square heads tried to suck my blood. I prayed for an end to the journey,

though absolutely none was in sight. This was one of the worst days of my life.

Finally, after several hours, trees appeared naked before blue sky on the crest of the hill in front of me. We had reached the top of a mountain. Below, a valley filled with grasslands and gardens gently sloped down. I clapped my hands as I spotted villages I had visited near Waisa—familiar lands, signs of human life. We had come out on the other side. I sped up my pace as the sun emerged, brightening the light green carpeting of grasses around us. Occasional trees rustled gently in the breeze.

Eventually we reached Waisa. Once home, I built a fire in the wood-burning stove using freshly cut logs. An orange blaze crackled, and the fire whipped into a frenzy. Maryanne then decided to bake bread in the hot oven. I stepped outside as the sun set behind the mountains over which I had just traveled. I was amazed I had made it. I both loved and hated this land, glad to be here, but aware I would be ready to leave as soon as my work was done.

The sun dove further behind the mountain, into rings of purple and honey-colored clouds. From the house I caught the warm smell of freshly baked bread. I turned around and entered.

Feasts

A few days later, to collect data I traveled back to Awande, where Carleton had once built a hospital. Few people were around, but I made arrangements to return two days afterward, and the villagers promised they'd gather elders for me to interview then. I was skeptical, but had little choice. Yet when I returned, the villagers, partly out of admiration for and gratitude to Carleton, had kept their word. A large group, including several elders, assembled.

They told me about two patients, Kanisi and Opaba, who had both recently died of kuru, and were brothers. Here, then, was a possible cluster of patients.

Opaba had fallen ill a few months before his brother and had died a year later. Kanisi had died six months afterward.

Beside me sat Amabi, whose mother-in-law, Neno, had been eaten in a cannibalistic feast where their relatives—Opaba and Kanisi—had been present.

"Were you at the funeral?" I asked. She nodded. As the ground was muddy, I had removed my poncho from my knapsack, partly unfolded it to make a small mat, and sat on it, cross-legged. As we had been sitting for a while, my legs hurt, and the muscles around my hips felt atrophied. The Fore, lacking chairs, could squat for hours, but I could sit in that position for only a few minutes before my weight squeezed my calves painfully, cutting off circulation. To sit cross-legged helped, but was still uncomfortable, given my heavy, mud-clad boot and my sore foot. But despite my discomfort, I was excited. Here was someone who had been at cannibalistic feasts in the past.

A large crowd began to gather around us. "What happened to Neno's body when she died?" I asked.

Amabi looked at me, surprised at my asking. "Katim na kukim na kaikai" (Pidgin English for, "cut up, cooked, and eaten"). "It was taim bilong kukum na kaikai" ("the time of cooking and eating"), the time of cannibalism, when the deceased routinely were consumed, a period clearly defined and demarcated from the present.

"What happened after that?"

"Putim long matmat" (bodies were "put" or buried in a grave).

"Why did cannibalism end?"

"Patrol officers came and told us to stop. Then Mr. James, the missionary, came to the Fore. Then Miss McGill came. Then the patrol officer put people away in *kalabus* [jail] for cannibalism. At first we went on in secret. Then, slowly, we ended it."

"Didn't you stop because of kuru? Because it spreads kuru?"

"No. It doesn't cause kuru. Sorcerers do."

Amabi told me that a few years before Neno's feast, Tononda, the boys' aunt—their father's sister—died of kuru. "But to get to her village required passing through lands belonging to enemy villages. Anyone traveling there risked being shot at with arrows and possibly killed." Only a few men dared trespass through this area to go to Tononda's funeral, and not all risked taking their wives and children. "But Tononda's brother, Atia—Opaba and Kanisi's father—went, and took his wife, Aiki, and their children. I went as well. Aiki sat closest to the corpse and helped cut it into large pieces and cook it in Tononda's garden. Next to Aiki were their three children—her daughter Agaro and her sons, Opaba," aged 8, "and Kanisi," aged 3. Following tradition, Aiki, as the sister-in-law, was handed the cooked brain, which she ate with her hands. She then did not wash her hands for several weeks afterward.

Then Neno had died—after the first patrol station had been built in Okapa (which I knew to be 1954), but before the road to Okapa was built (in 1955). Neno's husband, Kawata, was the village fight leader, a "big man." Many people attended the feast. I went down the list of sixteen of her kin: fifteen were in Awande at the time and fourteen were involved in the feast. The only one who didn't participate was Kawata's second wife, Masino, who had just married him from another area. "By tradition, she cannot eat the body of her husband's other wife," Amabi explained to me. "Here she is." Masino nodded to me—still alive, though elderly, and showing no evidence of kuru.

Of the participants, twelve later died of kuru. Amabi told me that the villagers had gathered in Neno's garden, as was the custom. The closest female relatives, including Aiki—Opaba and Kanisi's mother—sat closest to the corpse. Neno was the last person consumed in the village.

"Did you eat any?" I asked.

She nodded. "The hand," she said—a body part not likely to have contained any of the infectious agent, which is concentrated in the brain.

I went down the list of kuru patients from Awande. Two were at Tononda's feast. A few were from other lines or villages or born afterward. I didn't want my informants to get restless and tired, so I just asked about some of the other past patients. I asked about all the children who had died, and all the recent patients. Another boy, Ona, born one year after Kanisi, had died six years before Kanisi. Aosi, one year older than Opaba, had attended both feasts and died within a year after Kanisi.

The two brothers ate at both feasts. The other kuru patients had all attended other feasts as well. Opaba, Kanisi, Aosi, and Ona were thus all infected at one or both of these two feasts. The first three had almost identical incubation periods, while Ona's was six years less. *These were the only feasts the boys had ever attended.* Until this moment, here in this village, no multiple cases of an infectious protein disease had ever been traced back to specific one or two episodes of exposure. These periods of time were almost identical for the three children.

A man standing behind me spoke up. "I was at the feast, too," he said. "Though I was not a close relative, and ate only Neno's finger."

Another man yelled something out in *tok ples* from the back fringe of the crowd. Everyone laughed except me, since I didn't understand. Sana translated: "He said he ate a foot." I laughed too, and then everyone else laughed again because I had, after the translation. I pointed to my head.

"It's a good thing you didn't eat the brain," I said.

I interviewed them for another two hours, and then got up to go.

That these incubation periods could be identical in two or more people had never before been demonstrated. My efforts had uncovered a small truth—a tiny, hard, immutable fact about the agent and the way it works. If this phenomenon of long incubation periods could be better understood, infectious protein and other diseases could perhaps be prevented or treated. Yet these identical incubation periods here could also just be a coincidence. I would have to look for other such clusters, too.

Sorcerers

"I cure kuru," a short man named Sila told me a few days later. My guides and I had been hiking on a trail along a steep ridge. Far below, a lush valley spread.

"Wait," Sana had said. He called out into the valley in a high-pitched yodel. Banana leaves rose and tumbled over the underbrush. A few palms shot up, tall and slender, rare straight lines against the tangled landscape. After several minutes, plants near us began to rustle, and Sila had emerged through the thick undergrowth, having crept up the hill. My guides introduced us, evidently thinking that he, as their local medicine man, could instruct me how to treat kuru effectively. My guides now nodded, approving of his claims.

I squinted at him through the glaring sun. He was barefoot, and wore a soiled pair of shorts and an old Australian patrol hat. The hat's brim was unraveling, and had been fastened with a safety pin, which, though rusting, was the first one I had seen in the jungle. I was dubious about his remedies. What could I possibly learn from him?

"How well has your treatment for kuru worked?" I asked skeptically.

He listed his patients—Saroma, Saluta, and others—most of whom, he said, had recovered. In the course of my travels through the region I had met some of these successes and thought they seemed healthy. He admitted that several of his patients had worsened.

"Who?" I asked.

"Wanebi, Owa, Wasoru, . . ." All of them I had been taken to see, and had diagnosed with the disease. Our lists of current kuru patients were, to my amazement, identical.

"What's the treatment?" I asked, still suspicious. I wasn't sure he would tell me.

He did answer, saying that he first uttered an incantation, and then dispensed herbal medicines and prescribed several behavioral changes: for one week, patients were not allowed to drink water or eat any salt. As native salt was sodium–poor, the Fore usually ate as much of it as they could. "Lastly, for a week the patient is not allowed to touch anyone of the opposite sex," he added. The concept of "weeks" had only recently been introduced by Western missionaries and government officials.

Sila spoke with confidence and authority, which puzzled me. I knew that dozens of the Fore's herbal medicines had been tested for pharmaceutical activity at the NIH, and had been shown to be ineffective. But, I was beginning to realize, as the disease had wiped out most of the population, people who developed headaches, backaches, or a host of other minor pains now automatically feared they had kuru. These cases, which I labeled as misdiagnoses and as hypochondriacal, presumably constituted Sila's "cures," and lent credence to the Fore belief that the disease was caused—and could be cured—by magic. The tribespeople thus had an explanation they could understand for the horror in their midst, providing them with some sense of control over the epidemic. (Rarely, over the years, someone also developed malaria after working transiently on the coast—though there were no such cases when I was among the Fore—and such individuals were often thought to have kuru as well.)

Still, none of the patients whom I had diagnosed as having kuru had responded to Sila's therapy. Why, then, didn't he see a problem here, or feel troubled by the ineffectiveness of his medicine—the fact that some patients were still diseased, despite his treatment?

I didn't want to offend him, but I felt compelled to ask, "Why are some people whom you treat still sick?" I wondered how he would reply. Was it inappropriate to challenge an informants' beliefs head-on?

He didn't flinch. "Very simple," he replied, rolling his hand out in front of him. "These patients didn't follow my advice. They drank water or ate salt or touched a member of the opposite sex." In short, they were noncompliant. He blamed these failures not on the treatment, but on the patients. They had been bad patients, noncooperative. No one ever called the treatment itself into question. Even though some of his

clientele became sicker, the perception that other
demonstrated to the tribe that his approach could p
The numbers didn't dissuade anyone.

Sila explained to me how sorcerers in enemy villa

The sorcerer first stole a possession from his victim, then wrapped
it with leaves around a stone, cast a spell on it, and buried it in the
ground. The stone would start to rattle, and when it did, the victim
would begin to shake and get sick. Rocks claimed to be such kuru
stones had been dug up and displayed.

After kuru initially broke out, hamlets that had all been in one
area now spread out away from each other for added protection. When
deaths continued, the survivors moved even further apart, thinking
that more poison or sorcery must be nearby. When kuru continued anyway, tribespeople guarded their belongings more fiercely to prevent
theft. Yet still the deaths mounted. It was then rumored that a sorcerer
no longer needed to procure one of his victim's possessions to cause
disease, but could make do merely with bits of potato peeling the victim had handled, or chewed sugar cane. Hamlets erected stockades to
keep enemies out. Still, kuru raged. Neighboring groups that had once
intermarried with the Fore now stopped, fearing the Fore's dangerous
sorcery. The Fore were shunned.

When kuru deaths had continued, tribesmen then held large meetings, called *kibung,* to beg any sorcerers present to relent. Some atten-
dees rose, claimed they had been sorcerers, and agreed to stop. These
meetings were first held in 1957 in the North Fore area—a region entered by patrolmen and missionaries before the more distant South
Fore area. As a result of pressure from these Westerners, cannibalism
had ended sooner in the north. Shortly after the *kibung* were held, the
incidence of kuru decreased there. These meetings, credited with the
success, were subsequently held by the South Fore as well. However,
the disease continued in the south, since cannibalism had persisted
there longer. The South Fore then claimed that the sorcerers' promises
in these meetings had merely been false. Subsequently, the South Fore
guarded their belongings with heightened vigilance.

I had seen how widespread these fears remained.

One day, I had washed my clothes by hand and hung them out to
dry on a rope that stretched from the porch to the lemon tree. I had forgotten about the laundry when rain began to fall. Suddenly, I had heard

a knock at my door. It was Busakara. He had brought my laundry in. "It's no good someone takes it," he said.

"No," I answered, though I wasn't very worried about anyone stealing it.

"It's not good someone makes a kuru bundle out of it and uses it against you," he explained.

"Thank you," I said, taking my clothes from him.

Another time, Ganara, as I hiked with him, started chewing long stalks of sugar cane he had in his *bilum*. But he spit the pulp in his hand and kept walking, rather than drop it on the ground. "How come you're carrying it with you?" I asked.

"I don't want anyone to get a hold of it."

Similarly, to prevent anyone from making a kuru bundle against them, the Fore defecated in very deep latrines (surrounded by little *pit-pit* fences) so no one would procure the feces.

As I stood with Sila and the others now, I turned to them. "But kuru is caused by a tiny life-like thing," I explained. "A virus—smaller than an insect. Not sorcery."

"Show it to us."

"It is too small to see with your eye and requires a special instrument to view."

"What does it look like?"

"We aren't sure exactly."

"Has anyone ever seen it?"

"It's not clear anyone has."

They all laughed at me. "You White men don't make sense. Kuru is caused by kuru bundles, which we have all seen with our own eyes."

"But do you really believe that I could get kuru if, for example, a sorcerer took my boots?"

They all nodded.

"But you have only heard about the sorcery. There is no proof."

"No. We have seen the kuru bundles ourselves."

"But kuru is decreasing, because cannibalism is no longer practiced."

"No. Fewer die of the disease because the sorcerers have finally heard our pleas, and have seen the evil they have done. Besides, only the older generation knows the poison. The younger generation doesn't possess the knowledge."

"How come anyone dies then?" I asked, thinking I'd stump them.

"Because a few old timers are left here and there who still practice the sorcery."

"But small children no longer die of the disease," I said, summarizing the epidemiological data I, along with Carleton and Michael, had been finding. "Every year, the youngest people to die are older, since the last feasts are further and further in the past. The youngest patients used to be children, and are now adults."

"That's because children haven't lived long enough to anger the sorcerers," another man with us told me. "But others still die. My own brother died of kuru last year."

Sana and the others nodded.

Statistics didn't persuade people who had all lost close relatives to a disease. They all knew people who fell victim each year, and thus viewed the epidemic in personal terms, not abstract statistics. After all, even Sana and Sayuma believed in sorcery, though they had more exposure to Western science than anyone else here—working closely with leading researchers, including two future Nobel laureates.

I was frustrated at my inability to convince these people. Why couldn't they see that my views were correct? These barefoot villagers, born in a Stone Age society and lacking a written language, were challenging scientific truths and the scientific method itself—achievements considered to rank among the high points and defining accomplishments of Western culture. I was representing Western science and civilization, but couldn't defend them. These New Guineans readily refuted to their satisfaction, and dismissed the "universal truths" I had been taught and had accepted through years of Ivy League science courses. I couldn't disprove the Fore's arguments without first persuading these villagers that hypochondriacal cases existed, and that my definition and diagnostic criteria for the disease—rather than the Fore's—were accurate and should be followed.

Yet they had seen and known the disease intimately their entire lives, while I had just landed in their country several weeks before.

Sila now turned to me. "If you White men think you know what causes kuru," he asked, "why haven't you cured it? We've cured the disease. You haven't!" The natives all believed that Sila's treatment was effective, and that his theory was thus correct. In choosing between alternative, competing hypotheses, the Fore selected the theory that in their eyes produced a cure and therefore worked.

"But discoveries in the West take generations—even hundreds of years," I explained. The group wasn't impressed. "All the Western goods you see took a long time to invent. My watch," I continued, pointing as I spoke, "my truck, my boots all took many, many years to develop." But the Fore didn't have any knowledge of decades. The concept of a year (and hence of a calendar) had just recently been introduced by Westerners, and that was done only to mark the arrival of Christmas. To say "three years ago," one says *bifo tripela krismas*—literally "before three Christmases." Previously, the Fore didn't measure or chart the passage of time as we do in the West, and did not have a sense of time extending more than a few days forward or backward. In *tok ples* the word for "three days" ago was synonymous with "the past," while "three days from now" also meant "the future."

These native beliefs had led the people to take Western technology for granted. Twice on my travels in the bush I had passed wreckage of planes that had crashed and exploded during World War II. The natives had removed any smaller bits that could be of any use in daily Stone Age life, but the larger hunks remained—pieces of fuselages and broken pale grey wings of 1940s planes, scattered over quiet jungle mountainsides. These New Guineans must have feared and wondered at these large shining birds falling out of the sky in torrents of fire, smoke, and roar into this quiet world of stone and bamboo. The New Guineans, lacking engines, fuel, or any mode of transportation other than bare feet, hadn't even yet invented the wheel—except as part of a child's toy I happened to see. By joining two gourds together—poking into each the other's stem—boys created a pair of wheels connected by an axle that could be rolled along the ground with a long stick. Yet the potential of this invention had never been realized. The self-sufficient older tribesmen weren't interested.

The elders, when I asked them about the planes, invariably acted nonplussed. The New Guineans accepted the notion that White men had dropped from the sky in soaring birds. Shock is hard to sustain. Once, walking down a path with Sana and Sayuma, I pointed up to the full moon. "You know, we've sent men up there to walk on the moon," I said.

"Oh, yeah," Sana answered, not particularly interested or impressed.

"Really. They walked on the surface of the moon up there." I pointed again, at the far-off disk of white light, brighter in the clear sky of New Guinea than it ever was in the polluted air of New York. Sana and Sayuma nodded politely.

"Em stret," ("That's right") Sana said. He wasn't surprised or confused. His view of man and of Westerners hadn't changed. How could it possibly matter to him that we had gone to the moon? Perhaps he wondered why we bothered, how it could possibly concern us. My claim didn't disturb his beliefs about the universe, and probably only confirmed his notion that spirits determine how everyone fares in the world and that Westerners just have good *maselai*. His beliefs continued, supported rather than canceled, in his eyes.

Yet, the specific differences between our views often amazed me. On Independence Day—the date when Australia granted freedom to Papua New Guinea—the Lewises had planned a celebration at the local school. But when the day arrived, the natives almost all stayed in their houses. A few ventured to the school grounds, where they felt safe, but none dared go to their gardens or anywhere else. They thought that independence would unleash the spirits—that rocks and trees would shatter and be destroyed.

In fact, the Fore's cultural richness and imagination manifested itself most clearly in these elaborate beliefs about sorcery and magic. Such beliefs die hard.

Yet the coherence and strength of the Fore's logic impressed me. They hadn't discovered science, but then neither had the ancient Greeks or Romans. A few years later, I wandered through the ruins of Pompei, past ancient snack bars, brothels, theatres, cemeteries, ballfields, gardens, fountains, living rooms, dining rooms, kitchens, bedrooms, frescoes, mosaics, sculptures, furniture, graffiti, plumbing, and heating systems. What, I wondered, did we in the twentieth century have materially that they didn't? Electricity—and hence telephones, radio, television, and computers. Better machines. But not very much else. (Thanks to science, we also have better medications, such as antibiotics, which are extremely beneficial, though some ancients lived to be as old as anyone does now.)

Years after meeting Sila, as a young doctor, I would also hear arguments similar to his used to explain the successes or failures of treatments in American hospitals. Doctors blamed therapeutic failures on patients' poor compliance or low motivation, rather than questioning the efficacy or limitations of the treatments employed.

A week after my meeting with Sila, Wanebi, the first kuru patient I had seen, died. That night, her family held a divination ceremony was held to determine which hamlet the sorcerer came from. The mourners shook an arrow up and down in a bamboo tube, listing names of villages until the arrow popped up over the tube's rim. Then they listed the names of house lines until the arrow poked up again. As a result, the family would now avoid that hamlet altogether, or walk nearby only with another person. In the morning, they held a Christian service at the Lewises' church and had a Christian burial for her. They didn't see any contradiction between the two systems of belief.

A few days later, Sayuma took me to see the kuru stone said to have caused her death. A hard, brown, pockmarked rock, clumped with dirt, sticks, and leaves, sat in a flat clearing in the middle of a hamlet. The round, unearthed stone resembled a large egg. It now lay captured, surrounded. The rock was as inanimate as the kuru brains I had once seen at the NIH. Western scientists believed that a tiny protein inside the brain—not yet seen or understood—caused the disease. The Fore blamed the poison wrapped around this stone. The crowd pointed angrily at the rock, shaking their heads and mumbling to each other. I moved in closer along with everyone else to look. In the sunlight, the rock glowed, and for a moment, seemed to have an aura of its own.

Dinosaurs

Sayuma's son, Jason, knocked at the door one morning, holding a guitar. I didn't know where he had gotten it. The wooden instrument, scraped and held together at points with Scotch tape, bore a long history of survival in the face of overwhelming odds. I picked it up. It now had five—instead of six—old strings. I strummed. They vibrated dully. I tried tuning it. The pegs turned in their holes, but the strings slipped in and out of tune with the slightest touch. I played familiar chords, remembering melodies from far away.

"Play me a song," Jason said. I played fragments of "Alice's Restaurant" and a Bach melody.

"Play one for me," I said in turn, handing him the guitar. He grew bashful. "Please?" Jason finally agreed. He knew two chords, and positioned both nervously, his hands sweating on the fifth fret. I told him to finger them in the first fret, and taught him a simplified B flat, to complete a triad that could allow him to play a myriad of songs.

"I want to be in a rock band," he told me. He liked to sing, but was embarrassed to do so for me. He preferred to hear me. Still, a genuine warmth radiated between us. He talked about boxing, disco music, dancing with boys who were his friends at a Goroka disco. "What is a free country?" he asked.

"It is where there is freedom of speech," I said. He didn't understand. "People can say and do what they want."

"What happens in other countries?"

"Many are Communist, where people are supposed to share what they own." It was difficult explaining this in Pidgin. "But where the government prevents people from doing and saying what they want."

"I want to live in a free country," he said. I felt he didn't understand, and somehow saw himself as living in a country that wasn't "free." I showed him a map of Papua New Guinea. "I want to go to Mount Hagen, Port Moresby, and Rabaul"—an island off the coast. These were the limits of the known universe for him. He glanced at the pictures in a *National Geographic* magazine in the house. "What's that?" he asked, pointing to a painting of a dinosaur, scales thrusting up from its back.

"A dinosaur. They lived tens of millions of years ago."

"How do you know?"

"People have found their bones."

"What happened to them?"

"They died out."

"How come?"

"Nobody knows for certain. They may have eaten all the trees and then starved. Or the world may have changed around them."

"Did men starve then?"

"There weren't any men—not yet."

"What do you mean?"

"Men only were born in the past 100,000 years. But the earth goes back four billion years."

Jason had trouble understanding. "Miss McGill says there was Adam and Eve on the sixth day," he said.

I didn't want to contradict or embarrass Hazel. "There are people who believe that," I said. "But not everybody."

"Are you a Christian?" Jason asked me.

"I'm Jewish."

"Jew?" he asked, confused.

I couldn't begin to fathom what Hazel and the Lewises had said about us.

"Yes." He peered at me, including the top of my head—I had the feeling he was scanning for horns.

"Here are other animals," I said, returning to the book.

"What are these?" he asked.

"Fish." He had never seen any. "They live in the sea," I explained.

"What do they eat?"

"Often other fish."

"How come they live in the sea?"

"All life started there, and then moved to land."

"How come fish stayed?" he asked—partly curious, and partly testing my argument to see if it held up.

"They just did. Some, like dolphins, evolved from animals that lived on land and then went back to the sea." This confused him even more.

He looked at a pencilled self-portrait I had drawn using the tiny shaving mirror in the house. He then asked me to draw him. He sat well, though unsure of how he should look, never having seen a portrait, other than that of me. Like most boys here, a hole connected his two nostrils. In it, he and other boys put pencils—which fit—when not being used, just as Westerners insert pencils behind the ear. As I sketched him, he looked sad—it seemed at the limited opportunities for the rest of his life, and for his people. We got along, partly because I was younger than the other Westerners here, and because he had not grown up wholly in the Stone Age. I was reminded again of how you just get along with some people better than others, for whatever reason, even if you're from the West and living in a Stone Age village. In the drawing, I tried to capture his warmth, his mournful eyes, and the cruel wall between our worlds.

A week after Jason's visit, I saw Miss McGill. "Jason Sayuma stole some books from me," she told me.

"Really?"

"He asked to borrow them, said he'd return them the next day, and four days later he claims he had left them in front of my door."

I felt badly for him; I believed he was genuinely curious. He sensed the aura of books—that they held a certain power. In a more literate culture he might have thrived. But here, tasting one culture's treasures but caught in another, he only felt frustrated.

I wondered what kind of future he really had.

Meanwhile, word spread quickly about the drawings and others came to gape at them. The Fore had seen photos, but ironically never a drawing capturing a likeness and certainly never one made in their midst—something they might potentially do themselves. They had no representational art—none that strove after mimesis, which I had grown up with in the West. They had bodily decoration, squiggling colored dyes across their foreheads and cheeks and sticking feathers and flowers in their hair—referred to as *bilas* or ornament. Indeed, Highlanders had an almost innate aesthetic sense. Fashion appeared to be a basic human instinct. But the notion that art can capture the likeness of a specific individual was absent—it had not yet been discovered. This concept had, after all, only started in the West a thousand years before the Christian era—in fact, only at the end of the archaic period in Greece, several hundred years B.C., did sculpture become less stylized and more mimetic. Stone came to life.

Months later, when I eventually packed to depart from Waisa and return to the United States, the natives asked me to leave the two portraits there. I wrapped each sheet of paper in Saran Wrap for protection, and taped them to the bamboo wall. When I spoke to Mike Alpers a few years later, he told me that the pictures remained there, still viewed.

Wires

Roger needed to haul water tanks out from Goroka. On Friday, he got a ride there with the Lewises, and returned on Saturday with a loaded pickup truck that he borrowed for a few days from the Institute. "The next shipment of tanks comes into Goroka on Wednesday afternoon," he said. "I'll go in then. I also need diesel fuel. By the way, Michael is leaving on Wednesday for a few weeks." I decided to go to Goroka before that to talk to Alpers about what I had been finding thus far. On Sunday I took a PMV to Goroka. "I'll drive in again on Wednesday," Roger told me before I left, "and take you back out here."

"I came to town to see you," I told Michael when I arrived. "I gave myself two days, as I thought you might be busy the first." He seemed surprised. I had adopted his technique—to avoid being pressured by time.

I bought wine and New Zealand cheeses, and relaxed at his house, listening to Monteverdi choral works, Bach and Mozart, and Beethoven piano sonatas. Wendy, Michael's wife, was bright, enthusiastic, and interested in my research though I felt it must be hard for her, having visitors like myself come and go.

I waited until Monday morning to talk to Michael about work. When I finally sat down with him, I asked him about what to do in difficult situations with my guides, for instance, when they wanted to bring along extra carriers.

"They're merely trying you," he said. "Just say no."

I went on to tell him about the cluster I had found. He was excited and encouraging.

After my meeting, I went by the mailroom. A friend in New York had sent me a book to read—Antoine de Saint-Exupery's *Wind, Sand, and Stars*, about the author's experiences working and traveling in another far-off region, North Africa. I sat down and started to read it, inspired by Saint-Exupery's evocation of the strangeness and enchantment of living in a foreign land. A young New Zealand researcher, Carol, working in a lab, came by to get her mail. She had received a huge stack of letters. "That's quite a bit of mail," I said.

"I simply mail a lot of letters out. That way I get a lot in return."

She was tall, with long blond hair, and was wearing pants, topsiders, and a white blouse with the buttons—even the top ones near her throat—fastened. "What's it like living in Goroka?" I asked her.

"I've been living with a native," she said. "But they're so hard to understand. I told him to move out last night."

"How long have you been together?"

"A month."

"And you're living together?"

"Yes. But I've had enough." It somehow seemed an experiment. "Call me and we'll have dinner the next time you're in Goroka," she told me.

"Sure."

Michael's youngest son, Bradley, was home. Michael was also putting up Sayuma's son, Binabi, sending him to school in town, since the only school where Sayuma lived was run by the Lewises and was very religious.

Bradley wanted to play cricket. As Michael was too busy, the two boys showed me how. They had never played baseball (which I then explained) and I had never played cricket.

I then showed them some card tricks. Binabi in particular was very interested. I asked Wendy if I could help make dinner. She had two bookcases of cookbooks, racks of spices that I hadn't seen since leaving the United States, and a full set of cooking utensils. I read recipes for, and made lemon chicken, duchess potatoes, and a salad with an oil and lemon juice dressing (since there was no vinegar in town). I had also bought some cheese and crackers, as well as ice cream and cake for dessert.

After dinner, I wrote letters and mailed one to Carleton, summarizing my work to date and telling him that I had met Miss McGill.

That night, Bradley and Binabi wanted to go to the movies that were in town for the week—Superman and Bruce Lee films. "Can you take us?" Bradley asked Michael.

"I don't think so," Michael answered.

"I can take them," I volunteered.

Michael looked at me, astonished. "Do you really want to? You don't have to, you know."

"No," I said. "I don't mind." I wanted to be helpful, and besides, I hadn't seen a television or film in months (though it seemed longer). The boys cheered.

The theater was just a large room with folding chairs and benches lined up, and a white sheet for a screen in front. Except for us, the audience consisted entirely of barefoot natives. Many wore only shorts—no shirts—or leaves around their waists.

First, several minutes of commercials played, leaping casually across cultures. Wrigley's had filmed a commercial in Pidgin English for "PK Gum" that showed New Guineans happily popping the gum into their mouths. Several men were garbed in profuse ceremonial dress, with colorful paint coating their skin. From each man's nostrils hung a round shell that covered his entire lower face. Each man now lifted the shell, opened his mouth, and plopped in the gum. I laughed. No one else did: nothing surprised them about the scene. Another commercial showed cases of Coca Cola being parachuted down into the Highlands. As they fell from the sky, the voiceover talked about the arrival of cargo.

I have never seen a movie with a more illogical, nonexistent plot than the Bruce Lee film. The dialogue was insipid ("Oh yeah, well take that!!!" one character screamed to another who was already half dead). Extraneous characters were introduced just to be killed. People who had been beaten up came back just to be beaten up again. Characters were suddenly being killed in mass fighting, when they were supposed to be somewhere else. I would probably never go to see these films in the States. I wondered: had I come to New Guinea for this? Yet no one in the audience minded. Both films had action—kung fu fights, fist fights, sword fights, car chases, escapes. The films were in

English, which almost no one in the audience spoke. But that didn't matter either.

For twenty minutes, a planet exploded. Odd shapes trembled as the camera shook. A dam burst, an earthquake toppled buildings in California, a jet burst in mid-air, a helicopter went berserk. Crescendoing *Star Wars*-type music repeated over and over as the credits in blue neon zoomed forward and backward through outer space—impossible to read and giving me a headache. But this jumbled package of *Star Wars, Earthquake, Airport, The Towering Inferno,* and other disaster films enthralled New Guineans as completely as any Middle American audience. I realized how Hollywood films had indeed found the lowest common denominator, such that barefoot people from primitive tribes—who spoke no English—loved them just as much as twentieth-century Americans back home. These films were proving their ingenuity, transcending culture and education, selling themselves to feed a universal hunger. I treasured the books and high culture in the Alpers' house, for providing ways of understanding this very different country, and frameworks for seeing human beings in different cultures. But it was Sensaround that equalized and brought the two cultures together.

Wednesday arrived. Rain fell throughout the morning. Roger did not arrive all day. I speculated that he had wanted a day off, or that the storm had washed out the road in Waisa, or the car had broken down.

Thursday arrived. "Get a car and drive out," Kanaua, who was in charge of the Institute's vehicles, urged me.

"Yes, you must get a car," Sayuma, who had arrived, too, chimed in. "Something might have happened to Roger."

"I'll wait until tomorrow and see," I said. The car would be useful for my research—I could keep it in Waisa for a few weeks—but would be a lot of responsibility.

"No, no," Kanaua said. "I have to be back here tonight, and have to work Friday, taking care of the animals at the Institute. I can only go today." He was also supposed to be watching Binabi, who was staying at Michael's, as Michael and Wendy had now left. "If we leave this morning, I can get back by PMV tonight."

I felt enormous pressure from them, though I didn't know why. I didn't know how to drive a stick shift, but Kanaua showed me how in a cleared lot across from the Institute, and said he would drive me out. Finally, under pressure, I agreed to take a car.

But Kanaua didn't procure a vehicle for me until the afternoon.

"Can you take some cargo for me?" Sana, who had suddenly showed up, now asked.

"Okay," I said. I ended up bringing sixteen bags for his store. I was shocked—they wanted the car for themselves, not me!

Kanaua dropped off a note for Binabi at Michael's, and we left. We didn't get to Waisa until early evening. Roger was there—he had merely changed his plans, and decided to wait to travel to Goroka. Outside, when I finished unloading the jeep, I saw Kanaua sitting down to play cards. "Aren't you going back to Goroka?" I asked him.

"No, I think I'll stay here."

"But what about taking care of Binabi?"

"My note told him to take care of himself."

I was flabbergasted. "What about your work back in Goroka tomorrow?"

"I got someone else to take care of it," he said casually. I realized I could be blamed by Michael for having brought Kanaua here, knowing that the latter was supposed to be watching Binabi and had work to do. I had been set up—I just couldn't win in this culture. I was incensed. Kanaua had just wanted to leave Goroka for the weekend a day earlier and for free. I had been had; and was now stuck here with the truck.

"Did you bring any diesel out?" Roger asked me inside the house.

"No." He frowned. "I didn't know I should," I added.

"Now Roger's going to have to drive you around," Maryanne said. "You don't know how to drive a stick shift, do you?"

"I learned in Goroka," I said.

The next morning before dawn I left and drove off to another village with Sana and Sayuma. I headed from the main road onto a narrower one that was rarely traveled—it was simply a trail left by a bulldozer many years before. Rain started falling again, and soon began to gouge deep gullies into the dirt. Even a small amount of rain made these roads dangerous, if not impassable. Periodically the jeep got caught in quagmires of mud and we had to drag it out by hand. At one point the vehicle started sliding down a hill sideways. When we hit the bottom, I struggled to steer it at a right angle, and maneuvered it back onto the road. A mile further, mud trapped the car once more, and it again had to be pushed out. Some men walking along the road sat down to watch.

"Can you help us?" I turned to them and asked.

"Will you pay us?"

"No." I didn't have any extra money on me.

They continued to sit there. When we finally got the car out, they stood up, walked over, and asked for a lift. Shocked, I said I didn't have enough room. Their assumptions astounded me: they believed that Whites had an infinite amount of money and goods. They didn't think their own actions would influence my responses to them. I felt they didn't see me as human like themselves.

I drove on, but a few miles down the road a bridge had collapsed. I had to abandon the car and walk the remainder of the way. When we got back to the vehicle that night, the battery died after I started the motor. Without headlights or wipers, I drove through the darkening rain. A wet thick grey filled the air, obscuring both horizon and sky. I feared that if I shifted too quickly, the engine wouldn't restart. I sped to arrive at Waisa before nightfall; when I got home, I was exhausted and slept for ten or eleven hours.

I needed a day off, but patients and work awaited in every direction. The next morning I fixed the car's battery terminal. The wires had been weakly connected. The engine now started without a problem.

In the States, I never would have even thought of fixing a car myself. But here I had no choice. I had to learn the rudiments of repairing cars. Westerners were indeed specialized compared to New Guineans, each knowing all the technology in their culture.

In the morning, I drove to another, more distant hamlet. "Which is the best way to go?" I asked Sayuma. He pointed left. En route, he wanted me to stop so he could pick up one of his *wontoks*. I later learned that this route was far longer, but allowed him to give the *wontok* a lift. Sayuma felt no compunction about lying to me. The "truth" as I knew it—something final, absolute, and irrefutable—didn't exist here. With no written papers or documentation of anything, there was less verifiability or objectiveness about anything. Traditionally, Papua New Guinea had no written laws, history, or fact—only what could be passed on orally. The issue of which route would get me to my destination fastest was secondary to thoughts of whom I could drive where, or whom my guides wanted to visit. The notion of "the truth of the matter" was, I realized, wholly a cultural construct.

As I drove, we passed a man walking along the side of the road. "Stop," Sana told me.

"I am Committee," the man said. He was a local government representative. "Committee" had become his name. "You must give me a ride."

"You must give him a ride," Sayuma repeated.

"Okay," I said. Committee got in.

"You must give me a water tank," he said as we were driving along. "I am expecting one."

"Roger is installing the water tanks only in two villages," I explained. "Waisa and Purosa—as part of a study, to see how health and illness then change," I explained. This minister lived in Awande—a different area.

"But he wants a tank, too," Sayuma said.

We dropped the man off further down the road. When we arrived back at Waisa, Roger walked by. "The Committee from Awande needs a tank," Sayuma told him.

"I've told him 'no' many times," Roger answered.

"But he's a government minister, and wants one," Sayuma said.

"The tanks are only for people in Waisa and Purosa," Roger explained, annoyed now. "Because we want to understand how the tanks help people."

"That's what I told him," Sayuma said, now trying to please Roger and to appear to be the good citizen. "I told him he couldn't have one."

The Fore had no understanding of water purity. They wanted tanks not for cleaner water, but for the modern technology of the accompanying corrugated metal roofs—used to collect the rain water. The men were motivated by the status and novelty of ownership. Without science, they had different notions of what health, healthy food (they liked "grease"), or hygiene were. Older men wore on their faces layers of caked dirt—accumulated over years—as a venerable sign of age.

At the house, Bill and Terri Towson, who were both physicians working in Goroka, and their son Tim, about 3 years old, came to visit. Tim walked about constantly, sometimes with a pacifier still in his mouth. Terri paid little attention to him when he wandered outside, while she was inside, reading or talking. A PNG mother, in contrast, took her child—up to the age of 5 or 6—with her everywhere, suckling him or her.

At one point, Tim ran into Soba's garden next door, and trampled over vegetable plants. Soba saw Tim, ran out, and started yelling at Tim in *tok ples,* then in Pidgin, to get away. The boy removed his pacifier, put his finger to his lips and rubbed them up and down, mumbling, "Bububububububu." He then ran back into the house. Soba was incensed. Westerners easily bothered natives, I realized, as much as vice versa.

With the car, I was now able to travel further and could more readily return to villages for follow-up interviews. It was thus that I found a second cluster of patients in Ketabi village. When I had first hiked there, the villagers had all been at a funeral elsewhere. As a result, I didn't obtain much information about two recently deceased patients—Pikagu and Iyami. With the car, I could drive back to collect more information to see if they were related. I took Sana and Sayuma on the trip and we arrived at 11 a.m., giving me ample time. My guides spoke to several of the men, who brought out a few elderly women. The matriarchs marched out, self-possessed—the senior villagers. As I had lots of time, I was able to obtain a very full history.

It turned out that Pikagu and Iyami were both members of the same line, and had recently died within a year of each other. They both attended only one feast in their lifetimes—that of Pikagu's mother, Tomisi.

Taniya, an elderly woman, had been present at the funeral, and spoke with me along with two other older women. "I was not a member of their line," she told me. "I was at the feast but not very involved. Women and children gathered in Tomisi's garden to consume her. Her daughter Pikagu sat closest." Iyami, an infant, was two years old at the time. Twenty-eight years later, Pikagu developed symptoms in January, Iyami in July. I went down the list of everyone in the genealogy. Thirteen were women and children, and thus could have participated in the cannibalism. Of these, ten did, and eight of them later died of kuru. Of the two who didn't die of kuru, one died in childbirth. The other, an infant at the time named Iliru, was still alive, and also spoke to me. "I was there at the feast, too," he told me. "But I was too young to remember it."

The sun was still out. Since it didn't look like rain was near, and I had arrived earlier because of the truck, I decided to go down the entire list of the sixty-six patients from Ketabi and closely neighboring Ai and Purosa-Takai.

It turned out that forty-five people had gathered in Tomisi's garden that day. Twenty-one were from her hamlet, knew her well, mourned her passing most, and had seated themselves closest to the cut-up corpse. They would all go on to die of kuru at various times over the next twenty-eight years. Of the other twenty-one patients who were not present, seven resided elsewhere and only later married men from this area. Thirteen were born afterward. I checked the epidemiological record. Their years of birth were all later—through the 1950s. Moreover, none of the people born after the feast were said to have been there. I was able to figure out all of these years based on comparisons with the births of patients whose ages had been previously estimated by Carleton, Michael, and village census reports.

Of the fifty-six participants in Tomisi's feast, fifty-three later died of kuru. Two survived, and one died of something else. The incubation periods spread over a wide range. But Pikagu and Iyami, since there were no other feasts among their kin, had identical incubation periods of twenty-eight years.

Here in New Guinea, I was demonstrating the virulence of the infectious agent—that disease could result in humans from only one exposure at a meal—for the first time.

Years later, such facts and dates would be used to help calculate the upward estimates of how many people would die in Britain from Mad Cow disease. Yet these data also suggested that dozens of women and children—almost all those living in the village at the time—had attended each feast. Thus, most villagers had attended multiple feasts in their lifetimes, and possibly been exposed to the infectious agent multiple times.

I suddenly realized, looking around, that some old women had survived, even after all others in their generation had died of kuru. I wondered why. Taniya's son had succumbed to the disease, revealing that she, too, had been exposed at the feast, since mothers fed the brain to their children. I spoke to one of the other older women present, Anasuga, who was from another village and had not been at the feast, although she had lost four children to kuru. Yet she and two of her children were still alive. Did she have some antibody or genetic trait against the agent? If so, the information could be potentially lifesaving to others around the world. Yet no one had ever even been able to ask such questions before.

Moreover, years later it would be difficult to generalize from kuru's decimation to the fate of Great Britain's population. Estimates that two hundred thousand people per year in the United Kingdom will eventually succumb to Mad Cow disease are based in part on kuru having wiped out ninety percent of the women and two-third's of the total population in certain villages. But this prevalence resulted from multiple exposures over many years. Consequently, not everyone exposed to Mad Cow disease once would necessarily die. Unfortunately, the degree to which Britishers were exposed to infected cows is unknown. In any case, calculating death rates for Britain, which some have boldly undertaken, is clearly more risky than assumed—if not impossible—given such unknowns.

Grease

A few days later, I went to visit Ronald Monroe at his Mission. "Be careful," Sana warned me, "Em i man bilong greasim yu tru." I didn't know exactly what Sana meant, never having heard the term in Pidgin before, but it seemed to mean that he'd somehow really grease me up.

Ronald Monroe was the founder and principal of Tagowa High School—a large secondary school. As a college chaplain in the United States in the early 1960s, he had interceded with the CIA on behalf of a foreign student. His church, embarrassed, had then sent him off to New Guinea. Here in the middle of the jungle he had built an enormous complex that was part high school, part palace—a tropical paradise, constructed mostly out of dark brown woven bamboo. In the back of his residence stretched a wide terrace lined with potted plants—the only ones I had seen in the jungle—overlooking a waterfall that splashed down into a brook which Ronald had dammed to create a huge man-made lake, reflecting the surrounding mountains.

Ronald was a tall man with white hair wearing a Bahamas shirt and white pants. He escorted me to his private office, then excused himself and left by a rear door. He returned a few minutes later through another side door—as if there were secret passages. The scene was right out of the James Bond movie *Dr. No*, I thought. He then proceeded to tell me about his school and its accomplishments. The previous year the ex-Prime Minister had been the high school graduation speaker, flown in by helicopter. As he stood in front of the lake, a large crocodile swam up and opened its mouth. A young boy stepped out (the animal was man-made) to welcome the guest to the high school.

Ronald clearly had a flair for showmanship. At a fundraiser in California the year before, the guests had just seated themselves when half-naked New Guinea boys suddenly barged into the room and charged up the center aisle, shooting arrows that flew just inches over the guests' heads. When the boys arrived at the dais, they formed a semicircle and began singing church hymns in three-part harmony. The relieved would-be donors gave millions.

The schoolchildren had spent months in arts and crafts classes stringing nuts and beads into necklaces placed in banana-shaped clay containers as souvenirs at the fundraiser. Yet the Fore themselves had never made such necklaces before.

Missionaries did much for New Guinea, but they were not always liked. In the Seng valley of Western New Guinea, or Irian Jaya, in 1968, particularly arrogant missionaries had pressured the Yali people to accept Christianity. But the missionaries only clumsily understood the local language, and mistranslated the phrase, "to love one's neighbor." Instead of using the term *"Ok-nyt-tuk"* which means "to give from yourself," they said, *"Ok-nap-tuk"* which means, "to steal." They thus tried to force the people to accept Jesus Christ as the God of "Stealing." The natives became so incensed, that they killed and ate the missionaries. To show they were stronger than white men, the people then presented the fingers of the missionaries to other nearby villages.

Ronald asked me to come back to speak to the school. "How about next Tuesday?" he asked.

"Okay," I replied. "If I can't, I'll let you know." Before I left, he also asked me to see a boy who was sick.

"Do you think he has malaria?" Ronald asked me after I examined the boy.

"I don't know," I said.

"Do you think he might?"

"He might."

"Would you take him to the Okapa hospital?"

"Me?"

"Yes, you have a car here and know about medicine. It would help him."

"Well, I . . ."

"It would be so helpful to us all here." Ronald smiled broadly and put his arm on my shoulder.

"I . . ."

"Oh it would mean so much to us. I would be very grateful to you." Ronald kept insisting, and made it extremely difficult to say no. Under the pressure, I finally acquiesced.

By the time I left, rain had begun to fall. We were still driving in the night when the car abruptly died. I turned off the motor and got out, opened the hood and tinkered with the wires, shut the hood, and was able to get the car going again. On we drove through the darkening rain. Because of Ronald's request, the drive ended up taking an extra five hours—two hours to Okapa, and three hours to get home on the dark, muddy road. I later learned that Ronald had a fleet of cars and drivers at the school.

When I got back, Sana met me. "Yu lukim long Ronald?" Sana asked (meaning, "Did you 'look at' or see Ronald?").

"Yes," I said cautiously.

"Em i man bilong greasim yu tru?" I now knew what the term meant ("Is he a man who greases you up?").

"Em stret," I said. "Em stret."

Cargo

"Our chickens keep getting stolen," Jake Lewis told me one day. "We still think Jason is taking them."

"No one else?"

"He was seen taking them last time."

I couldn't help thinking of the difference in economic status between the Lewises and the surrounding Fore. Yet when I next saw Jason, I asked him about the chickens. "I don't know anything about it," he said. He then grew silent and depressed.

I thought the Lewises might not like Jason simply because he wouldn't fully adopt their beliefs. "People will sometimes accuse you of things which are wrong," I said. "If you didn't steal the chickens and know in your heart you are right, then you should tell the Lewises. If, on the other hand, you did steal the chickens, I'm sure the Lewises will respect you more if you tell them and say you are sorry." I felt like a father. Dealings in such moral terms didn't seem to occur here, however. Instead, if the conflict persisted, opponents either fought or went their separate ways in the bush. Recently I had seen Sayuma, Busakara and others heading off with bows and arrows to a neighboring village. A man there had "touched" a woman he shouldn't have.

A week later, I drove back from Goroka with some supplies and a can of diesel fuel. Jason came out to help me unload the truck. I carried some parcels into the house. When I returned to the truck, he had disappeared. I reached in the back for the kerosene, but the can was gone. I stood perplexed, then suddenly realized that I had been fooled. Jason had taken it. The natives were skilled at getting what they wanted. In this regard, civilization had not advanced very far. The West has romanticized innocent savages. Perhaps, as soon as the West begins to contact them, their "innocence" changes. But I suspected the innocence was illusory from the start. They, too, were greedy, scheming, and deceptive when necessary. That did not, however, diminish the need to fight against these inclinations in daily interactions.

My guides and I agreed to leave at 7:00 the next morning. But by 7:45, no one had yet appeared. I started the motor, hoping to wake them. The truck sat idling for several minutes. I drove a bit ahead on the road and waited another hour. Sayuma and Ganara (who was staying in Waisa) still failed to arrive. Uncertain what to do, I left and met Sana further down the road, as planned. We drove down to a fork off the Goroka road, past Awande. No cars had driven on this stretch in a long time, and foot-deep mud stretched for miles. In first gear with four-wheel drive, the Land Rover rumbled slowly. Steadily, I pushed through the bogs. Finally, I parked and we walked up to the village, which sat atop a hill. On either side, valleys spread out, vast and green. I gathered information on a patient, Nilaya, as the rain began to soak us. Her genealogy was sparse, revealing no other cases of kuru, either recent or in the more distant past. I wanted to collect information on another patient—Sabamo. "You must talk to her husband, Simia," I was told by a man in the village. "He is in Yagusa."

"Oh," I said. I would now have to trek there as well! I sighed, overwhelmed by the prospect.

"Wait," the man said. "We will call out to him." The man rose, walked to the edge of the hamlet, on a crest overlooking a deep valley, and hollered several words in a singsong voice. Then he motioned for us to sit. An hour later, Simia cut across the bush and arrived. I examined his wife, who clearly had kuru. He also told me that his wife's brother, Senlado, had kuru as well, but had left the area and gone to Mount Hagen, located in the far Western Highlands toward the Indonesian border.

"Do you know how to find him there?" I asked.

"Yes, he is preaching there." Perhaps I could go to see him.

On the way back, I met Ganara on the road near Yasubi. He was sweating badly and looked downcast. I explained as kindly as I could that I was not cross, but that in the future he would have to come on time. It turns out that he had arrived at the house inquiring as to my whereabouts at 9:00 a.m., two hours after he said he would.

The following day, I wanted to go to Kanigatasa to see another patient. I suggested we leave at 8:00 a.m. But Sayuma said, no let's go at 7:00. Ganara arrived forty-five minutes early. Sayuma arrived at 8:00, and walked far ahead with Ganara and a friend they met on the road. They paid little heed to the growing distance.

A week later Sayuma said he wanted to come by to talk with me. I told him to stop by at 2:00 in the afternoon. He arrived at 7:30 p.m.— knowing I would be starting to eat dinner. Maryanne was already seated when he barged into the small house and took a chair near the table.

"What are you doing here?" I asked him.

"You told me to come."

"I said earlier."

"I am here now. I have several things I need to discuss with you."

"Let's talk about it tomorrow."

"No, we need to talk about it today. We work together, do we not?" He started to list patients.

"But we've gone through all this already. Why don't you come back?"

"Carleton and Michael would never tell me to leave. We work together."

I didn't feel comfortable demanding that he get out, but he obviously wasn't getting the hint. I realized how much I took such subtle cues and unspoken gestures for granted. They were culturally specific. Yet he was, after all, officially my guide.

"Okay, what is it?" I asked him.

"I want you to go to Goroka to buy me boots."

"You should go yourself to try them on."

"But you need to be there to pay."

"I don't have money to buy you boots."

"But I need them because of the work I do with you."

A few weeks later we went to Goroka. This time, Sayuma asked Michael about the boots. Michael said okay, and that I should take Sana and Sayuma to buy them, which I did. Sayuma insisted on buying size nine, though he had worn eight last time and the size nine pair was too big when he tried them on. But he believed that more was somehow better.

Two days later, he informed Wendy that he wanted her to take the boots back.

He told her that they didn't have size eights when he went, but that they said they'd have them in the next day. All this was a lie, of course—I had been there, and knew that they had size eights and that Sana had in fact gotten a pair.

"Why does he do this only to me?" she complained

"Don't worry," I assured her. "You're not alone."

A few days later in Waisa, Ganara came by, shuffling. He repeated that he wanted an Institute house in Paigatasa. I told him he'd have to speak about that with Michael. Still he lingered. I could tell he wanted to ask something else. "Well, can you hire some of my *wontoks* to help in your patrols?" he said finally.

"I don't need them," I said. "If I do, I'll let you know."

"Well, can you take my wife with an injured leg to Goroka?"

I wasn't sure how to answer, not particularly wanting to make the trip. Roger spoke up. "There's an ambulance at the Okapa *haus sik*" (from the English, "house of the sick," meaning "hospital"), Roger said, "that's just for such situations."

Ganara walked away sadly, not having gotten anything.

This society had many capitalists, but few ideals. Natives spoke of spirits, not ideals. Traditionally, this society was a stable, closed system that didn't require larger organizing principles as we knew them in the West. My guides worked not for the science, but for profit. They pushed to get as many material goods as they could for themselves and their *wontoks*. Ganara saw no point in what we were doing, other than earning him money. In fact, Sayuma was persistently said to be one of the leading sorcerers in the area. I wondered whether he was doing something that fueled this rumor. I felt drained from fighting against the ignorance, opportunism, and foolish superstitions. Yet I was dependent on my guides and translators.

Still, the ability of those raised in a primitive society to interact with a more advanced one—and vice versa—challenged the notion that men's thoughts or beliefs have progressed much over the centuries. Otherwise, how could an individual readily jump across a culture gap of thousands of years? The bases of psychological manipulation haven't altered. Cultures differ in how they approach essential human needs, but the basic psychological types of individuals I encountered here resembled those in New York. I was seeing firsthand how little human nature differs between societies.

Missions

"I need rice for my store," Sana told me the next day. "I want you to buy it for me and to drive my cargo home with me."

"I will let you buy your cargo, but only after the day's work is completed," I countered. Unfortunately, the quality of my guides' work did not merit the quantity of their requests. But they persisted in asking, testing my generosity and will.

Jason stood on the road, uncertain whether he was coming along or not. I hadn't seen him since he had stolen my jug of kerosene from the truck. "Can I come?" he finally asked. His eyes pleaded apologetically. "Sorry" did not exist here, though I could tell he wanted me to forgive him.

I was now short a man. "Okay," I said at the last minute. He grabbed his bag and came running.

I also gave a lift to Soba and a young boy he thought might have malaria, taking them to the Okapa *haus sik*. I had never talked to Soba by myself—he'd always dealt with Roger. Since he saw me as Roger's *wontok*, Soba avoided me. On the trip now, I was able to ask Soba about his sister-in-law, Tasa, who reportedly had died of kuru a few weeks before my arrival. He reported that at first she'd had a headache and some difficulty with all of her limbs. Soba thought it might be malaria, and got her chloroquine and aspirin. The medication didn't help. A few months before she died, he told her that he thought she might have kuru. She said that she thought so, too. During the last two months of her life, she "lay down pinis" (lay down finished, or finally, that is, for the last time). Soba went to see her, but she was cross with him, and said that she thought he was trying to "fightim" her. After that he didn't help her much and she died soon afterward. If I didn't see patients it was unclear what, by Western standards, they actually died of.

At Tagowa, I sent Jason to Ronald with a letter, asking to postpone my lecture on Tuesday, since I'd be gone. As there was no mail or phone service, all messages had to be hand-delivered. Jason went reluctantly. I could only hope he would deliver the letter.

I finally drove up to the village that was my destination—situated on the road closer to Goroka and hence more contacted by the outside. The local village court judge eagerly helped me, sending word out to a recently deceased patient's *wontoks*, telling them to meet me at the top of the mountain, in a clearing on the trail. "Thank you," I said. He smiled broadly.

"Now drive your car up, turn it around, and park it here in front of my house," he said.

"No, that's okay," I said. I had already parked my car just below the village, which seemed close enough.

"No, you must! You have to!"

"Why?"

"It is better that you park it here." Maybe I was failing to grasp some disadvantage of leaving it in the road, I thought. I didn't want to offend him—he had been helpful. I took him up on his offer, though slightly suspicious of his eagerness, and then conducted the interview. The patient had died of kuru the year before. I thought she might be related to a man who died of kuru at the same time in Awande. But her family was from one of the two hamlets in Awande, and the man came from the other—a tenuous connection. They had no family in common. It is unlikely that they developed kuru from the same feast.

I was about to leave the compound when the judge pulled me aside. "I want you to tow my car to Okapa," he said. "The tires have no tread."

"But I have other patients I need to see today."

"Then I will drive on my own, but you must stay just a little ahead of me. And," he laughed, "you can pull me out of the mud if I get stuck." I hesitated, not wanting to get too involved. "Okay," he said, sensing my hesitation, "just drive to the top of the first hill, and wait to see how I do."

"Alright," I said. After all, I didn't want to be ungenerous. I drove to the top of the rise. "He's stuck in mud," Sana told me, looking behind. I drove back down. The judge fastened a thin nylon rope to the two vehicles. As I'd expected, it snapped. They then fetched a weighty metal chain which yanked at my chassis, but blasted his car to the top of the hill. The chain was then unfastened. I had my three assistants clamber on board, and I sailed off briskly. After all, I had been asked only to tow him to the top of the hill, and had made no further commitment. I had work to do, I reasoned—I was not a professional tower, and besides, I hadn't asked him for any significant favor.

I sped on to Okapa. Sana went to buy bulk cargo at the store. Prices were thirty to two hundred percent higher than in Goroka. He asked me to give him money, but I declined. I told him I had funds in Goroka, but only a limited amount of money out here in the bush. He didn't appear to mind my refusal. He bought biscuits, dripping, and only one bag of rice, all of which cost 93 kina. "I could buy two times that in Goroka," he told me.

We headed to the Purosa road *bung* (intersection), arriving there at 2:15—too late to drive all the way to Purosa and perform any significant work, especially to see the current kuru patient. "Just drive me down to the rock," Sana said—referring to a large grey boulder, a landmark on the road. Yet when we arrived there, it was clear that Sana had too much to carry. I drove him on to his village—an extra two hours of driving in all. How did I end up in these situations?

On returning home, I passed a Waisa man, Ipaki, walking with his two wives and two children, and a man carrying a child whose foot was bundled. I stopped to give them a ride, and at the same time gathered some heavy stones to improve the traction of the tires on the road close to Waisa. The man said he was taking the injured infant, whom he was carrying in his arms, to the *haus sik*, and could get a lift from Waisa. But after I arrived at Waisa the man went to hang out on the sidelines of a card game, and was still there an hour later when I walked by. Ipaki and his line disappeared, and I ended up emptying the rocks out of the car by myself. Everywhere I went, I felt I was being used.

Reluctantly, after my last interaction with Ronald, I returned to Tagowa High School. I didn't think I'd get much out of the talk, but decided to go out of a sense of civic and scientific responsibility. I told Sana and Sayuma they didn't need to come, but they insisted on joining me. When I arrived, Jane, a young Californian, gave me a tour. She was a missionary there for a year, and had a pale face and a page-boy haircut. She played guitar, as I did, and was friendly. "The school has doubled to two hundred students, and is one of the few high schools in the country," she told me as she showed me around the buildings. In the end, we reached the main entrance. "There," she said, pointing, "is a store belonging to a Seventh Day Adventist. He is not a real Christian." The man, a native, had recently built a snack bar next to the property. "He had been the headmaster, but was no longer qualified," Jane said. "He became angry and threatened to make trouble by building his store. But students aren't permitted to leave the school premises. So, his plan will fail!"

"What does he do that's so bad?" I asked.

"He believes that Satan will come back to earth," she explained. "He lives by rules, not grace. He is Satan-fearing rather than God-loving."

"But what has he done exactly?"

"He believes in the devil."

"But how does that show itself?"

"His is a Satan-oriented religion." I saw I wasn't going to get very far with her.

For lunch we sat outdoors on a screened-in veranda. Ronald saw I had three guides. "I'll take them to the servants' mess," he said. He returned and sat me on his right. Uniformed servants brought chili and rice and cakes and muffins, and laid these out on a beautiful, deep navy-blue table cloth.

"It's ironic," I said, "that the American dream differs so much from the reality. New Guineans think everything comes easy there, and don't realize the hard work involved." As I talked, I felt very much at ease. It was nice to be with Americans again, even if they were missionaries. "By the way, what should I talk to the students about?" I asked Ronald.

"We like to have visitors talk about where they're from, and the history and customs of the place. The topics that interest the students most are dating and cars. But talk about religious institutions, too. Also speak about the facts concerning kuru. Students come to the infirmary with headaches or stomach aches, afraid they have it." Ronald told me about other kuru patients, too, about whom my guides didn't know. "Families still don't like to talk about loved ones who have the disease," Ronald explained to me. "Shame remains attached." For the names of these other patients alone, it was worth coming to the High School, I decided—an important part of my field work.

After lunch, I stood at the podium beneath a large thatched roof supported by poles like a tent. The entire student body of two hundred showed up, in addition to the whole staff. Around the sides of this open, unwalled structure stretched luminous green lawns, and Ronald's man-made lake. Thick virgin forest covered the mountains above us. To help them better understand America, I drew analogies to New Guinea. "The two are in some ways similar," I said. "America was once owned and inhabited by the Indians before White men came. Early cities like New York were ports just like Madang, Lae, and Moresby. Some early immigrants came as missionaries or, like the Pilgrims, for religious freedom. But Europeans fought the Indians and bought or took over the land. We, too, were an English colony, and became independent in 1776—exactly 200 years before Papua New Guinea. The opening of the Erie Canal—like the Highlands Highway— spurred development. Outsiders came for opportunities. New immigrants also spoke different languages—making America a land of many tongues, as is PNG. Attempts were made to protect beautiful areas—as the Fore do—and this is an important goal."

"In the last century, gold was discovered on the West Coast of the United States in California, just as in New Guinea—which led some prospectors to move further into those areas. Oil and other mineral resources were also discovered—like the mines here. Important civic projects were also then built: schools, museums, libraries. After World War II, African-Americans, and more recently Puerto Ricans, moved to New York, leading to a mix whereby people of different races—Black and White—live together in the city. New York has tall buildings, over 100 stories high, because land is scarce and tall buildings are thus efficient. Most people in the area live in suburbs, however, with grass and trees."

"Teenagers rarely own cars. More often, parents buy and let teenagers use them, but that varies with locale and is less true in cities, such as New York. Most schools are half boys and half girls." (Here in New Guinea, the high school was overwhelmingly male.) "Most people go to public school, some to private or parochial school. Nationwide, about half of people go to college."

"There are different religions, but most people have some religious instruction." It seemed odd to reduce America to this. "There are Protestants, Catholics, Muslims, and Jews."

"People in the US don't know much about PNG." This surprised my audience.

Yet it was kuru that interested them most—a topic they knew well. "Kuru is caused not by sorcery, but by a virus spread at cannibalistic feasts," I said. "Children used to get it, but now only adults do. The North Fore stopped the feasts first, and thus have fewer cases today than the South Fore."

I explained as much as I could, based on what I knew and was finding out. Then I asked to see a show of hands—how many people still believed kuru is caused by sorcery? Almost all hands went up. Then I asked how many people now believed it was caused by a virus. Only a few hands went up—mostly teachers, though not even all of them. I told them that no one I had seen with kuru had been born after 1960.

They didn't believe me, and a heated discussion broke out. One teacher, who had raised his hand the first, but not the second time, said, "you have to understand that the children here grew up with these beliefs. It is hard for them to think differently." It was true for adults, too, I realized.

"But you have only heard about sorcerer's poison," I argued, "while I could show you a photo of the virus" (actually of the protein clumps believed by most to be the infectious agent). "If I told you that by snapping my fingers you would all die, would you believe me?" I asked. "After all, I have only told you that. It has not been proven."

I realized that the logical implications of my line of argument were that one should only believe what one saw—that one had to see to believe. Extended logically, this scientific principle was inherently antireligious: one should not believe in a God one cannot see. Afterward, I mentioned this paradox to Jane.

"I realized that," she said. "But it's all right because you were only speaking in reference to kuru sorcery—an evil superstition."

During my lecture, Sana and Sayuma sat impatient and bored. But it was strangely exciting to fight superstitions about the disease, testing and seeing the power of myths, seeing these ideas fight for their survival. Ironically, I was upholding the cause of the scientific method versus mythology—at a missionary high school. But in the end, I had swayed some students. They had lived in constant fear they might die of the disease, and were now, I hoped, relieved. I also saw how tenacious these beliefs remained.

Tribal Warfare

"I think Jason is out there in front of the door looking for you," Roger told me a few days later. I stepped outside and saw Jason skulking around the house.

"What's up?" I asked.

"I'm hoping to get something round . . . with a hole in it."

I thought. "You mean a kina?"

He nodded.

"Why?"

"Because I have nothing to eat."

"What do you mean?"

"Sayuma won't give me any food nor money."

"Why?"

"Because I shot Binabi."

"You shot him?"

"With an arrow."

"Why?"

"He made me mad." I suspected Jason was jealous of the advantages his brother was receiving, living at Michael's. I gave him some rice, and he left. Half an hour later, Sayuma came by.

"Yes?"

"Jason wants some fish to go with his rice."

His audacity astonished me. "Go buy it at the store. You have money from the Institute."

"But Ipaki's store is closed."

"So go to the mission store.

He saw I wasn't going to give it to him.

"Alright," he finally said, starting to leave. I knew he wasn't going to go to the store.

"Besides, you have a garden," I added.

"But I don't have any food in it because a pig broke into it while we were out on patrol."

I shut the door behind him. "He's a walking shit," Roger said, having overheard the conversation. "Even if a pig did break into his garden, he has four gardens!"

The Stone Age produced scoundrels and rascals on the one hand, and gentlemen on the other—similar to modern civilization. I was coming to understand why capitalism worked as effectively as it did in the West, while its main rival over the past hundred years—Communism—lost out in the end. Capitalism followed innate human nature, which included competitiveness, ambition, and greed.

The next day I headed to Henganofi to see a patient. The night before, Sayuma had advised taking the Nuperu road, which looked like the most direct route on the map. But in the morning he reversed himself and told me to go via Kainantu.

"How come?" I said.

"The road from Nuperu isn't in good condition."

"Do PMVs take it?" I asked. He hesitated. "Well . . . Do they?"

Finally, after twenty minutes of vain attempts to clarify this point, I learned that they did. "Is it like the Purosa road or the Oma-Kasoru road?" I wanted to know. The former, I was told, which I knew to be in better shape than the latter. I persevered partly because I suspected that Sayuma just wanted me to drive him to Kainantu. I took the Nuperu road, which turned out to be fine, but at one point the bridge had collapsed. If I had gone via Kainantu, I would have gone back this way and then had to make a U-turn, and return via Kainantu—which would have been much longer, and hard on my limited supply of fuel here in the bush.

I parked as close as I could to the village. I would have to hike the rest of the way. Luckily, there was a short-cut. Dozens of children at a school watched me walk across the river barefoot. We continued up a steep mountain, Sana and Sayuma complaining as we climbed. "Maybe I should just take younger boys next time to help me," I said.

"No monkies em i inap pinis wok bilong mi," Sayuma said ("No children are able to finish or do our job").

"Em stret," Sana said. It struck me as odd that the Fore used the term "monkey" to refer to children. These tribesmen didn't know the prejudices associated with the term, but just accepted it in Pidgin, unaware of the meaning of the word—that it referred to a dark, subhuman primate.

"Monkies em i no got save," Sayuma added ("They don't have knowledge").

"Em stret," Sana added, nodding vehemently.

"Monkies can do plenty," I retorted.

"No gat," Sayuma said. "Em i no got save long face bilong man na merri" ("No, they don't have knowledge of men's and women's faces"). It was a clever argument—that Sana and Sayuma were needed because they were older, and knew more people.

Still, that was not what I depended on the two men for.

With the exception of Ganara in his area, my guides could potentially be replaced with more appreciative and hard-working younger assistants, for whom personal material advancement would be less of a concern, without sacrificing efficiency. Younger assistants would also be better satisfied with the pay. These men had a ridiculously inflated view of their worth to the work.

The trip to this village was arduous, and seemed even longer because of these on-going arguments. But when we arrived, the patient's oldest brother said, "No, you can't see her today."

"Okay, how about next week?" I asked.

"Okay," the younger brother said.

We left. A group of friendly children, thrilled to see a White man, followed us out. Two days later I drove to another nearby village. Sayuma saw and spoke to the brothers, but I decided to stick with the plan to see the patient the following week.

But a week later when we returned, the youngest brother turned me down again.

I got mad.

"I told Sayuma," he said. But no one had told me. I walked up to the patient's brother. "I only have a few minutes and want to look at the patient now," I said, "since you said last time that I could. All right? It will only take a second." I was sick of these games and needed to know if she indeed had kuru. None of the brothers said no.

I stood and strut smoothly, rapidly to the door of the hut. No one stirred. I knocked, said, "open the door," and entered. The patient had a mild tremor and difficulty sitting up—clearly early kuru. I was refused a family history. I thought of returning another time, when the oldest, sternest of the three brothers wouldn't be present. But how could the wonderful, innocent children who followed us home the previous week turn into such fearful and foolish adults?

M aryanne left for Goroka for a few days. Roger remained, soon separated from his wife for a week—more than ever before in his six-year marriage. Each day he waited anxiously for her to return home. He did not know why she hadn't yet. There was no phone or mail or telegraph. Wireless communication would arrive here, I imagined, before that using wires.

He was increasingly frustrated and on edge, and became antagonistic. One afternoon when I returned home from hiking, he stomped outside, the door slamming behind him. He started yelling at Soba about the latter's pigs running around our house.

The next morning, Soba walked up to me. "Roger is a no good man," he said. "He has a hot temper and got angry at me yesterday." Anger made the Fore uneasy. "I am going to talk to Michael and tell him and complain," he continued, "and the Institute house can no longer stay on my property."

"Look," I said, "I'm sure it'll work out okay. I'll speak to Roger. Why don't you just wait and see how it works out? I'm sure Roger did not mean to get angry at you."

I was sympathetic. Indeed, patience seemed called for: we were, after all, trying to bridge several thousand years. Soba felt insulted and hurt. He did not know the push and shove of the modern business world. In the West, we have erected masks—public defenses and disguises. But the Fore had had no need to do that. The only one who tried was Sayuma, who, gratefully, was usually obvious in his attempts. Soba, a grown man, was still visibly upset by the encounter.

"Just think how difficult and frustrating it would be for you or another Fore man to go live in America or Australia," I explained. "So, too, it is difficult for an American or Australian to come here."

He remained bitter.

"Just remember," I added, "you get many things from the Institute. You get paid kina to have the house here, and often free rides to

Goroka and elsewhere." This last argument quieted him. I tried to be as careful as I could around Roger, and made dinner more often, but he stayed irritable. Two days later, Maryanne returned.

A couple of weeks afterward, I went into Goroka to take a break, and to speak with Michael about the possibility of traveling to Hagen. I also picked Sayuma's pay for him. The day I returned, Sayuma came by in the morning. I gave him his money on the porch, but he then marched right into the house. I followed.

"What is it Sayuma?" Maryanne asked him

"I want a ride to Goroka."

"I don't think that will be possible."

He walked right past her. Fore men didn't respect women. The culture was very male-chauvinistic: men could marry as many women as they could afford to buy, and counted their wives along with their pigs as their chief possessions. Fore men also considered women unclean— prohibiting them, for example, even from walking over an area where food was laid. Men also feared women, who were said in Fore mythology to have once possessed the sacred flutes that the men had stolen, and now forbade women to see.

Maryanne stood frozen at Sayuma's effrontery.

"I need a ride to Goroka," he turned and announced to me.

Roger, still in bed in the other room, yelled, "No. Tell him no."

I turned to Sayuma. "No," I said firmly.

Sayuma shook his head and stomped out. Roger emerged from his room. "Why did you let him in?" he scolded Maryanne.

"I didn't say he could come in."

"Don't let him disrespect you like that."

"What did you want me to do?" she asked.

Roger turned to me. "This is all your fault," he said angrily.

"Look," I said, "Sayuma is the kuru worker that the Institute has here."

Everyone cooled off, but the underlying tension still simmered. A few days later I got back to Waisa after an all-day hike with Sana and Sayuma. We were exhausted from trekking, and I invited them into the house for a cold drink. "How was your trip?" Roger asked with disgust, foaming at the mouth.

"Okay," I said. Roger stood over us in the low-ceilinged room, frowning the whole time. Sana felt very ill at ease.

"Well, thank you once again," I quickly told my guides, standing up to indicate they should leave. "I'll see you tomorrow."

Sana understood and stood up. Sayuma kept sitting, wanting another drink.

Sana turned to him. "Let's go, Sayuma." Still, Sayuma sat there. Finally, Sana said something in *tok ples*. Sayuma turned around and left.

Roger shut the door behind them. "Sit down," he barked at me. I hesitated, but did as he asked. "You, too, Maryanne," he said. "I want you to know," Roger sneered, "that my work has been interrupted because of you and your guides. Michael doesn't care about our project. He cares more about yours. Your work is going well at the expense of mine!"

"That's not true," I said. "Michael cares about your work. He just doesn't show support very visibly. And my project has its own problems, believe me."

"Well, we get drained every day here by petty concerns. And now we have the car here, making things even worse for us. I get bugged by requests for lifts all the time."

"But the car helps you, too."

"We did without one before."

"Just tell people you can't give them lifts, and that I'm responsible for the truck, not you, and that they should talk to me about it."

"They know better. They see me drive it."

"Then just tell them 'no.' " Unfortunately, he kept saying yes and then resenting it.

"And you have Sayuma here all the time," he said.

"You think I want him here?" I fired back. "He pisses me off as much as he does you. Besides, he's not here 'all the time.'"

"All I'm saying is that we have no privacy. We're used to a very quiet life in Australia. Our only friends there were our parents. They're the only people we ever visited or were visited by. Now, we're in the center ring all the time. Today you took over the whole house for thirty minutes."

"Look, I'm sorry," I said, even though I felt I'd done nothing wrong. I was simply an easy target for all of Roger's frustrations. In fact, we were all constantly watched by the natives. "It's just really hard doing work here," I added.

"Don't talk to me about hard work," said Roger. "We've worked every day for thirteen months, without one day off!"

"Well, maybe you two should take a vacation for a little while."

"We can't. If we did, we'd never come back. We hate it here. These natives are all so fucking unappreciative! They're horrible."

"They're trying." Roger frowned; his eyes shot daggers at me. "Well, I'm going to go away for awhile anyway," I said. With that, I went to the outhouse. Built of pine wood, it was dark, quiet, and calm.

That night, in my room by myself, I almost cried in my pillow. I'd never felt as unappreciated in my entire life, and never so alone—thousands of miles from anyone I knew. I'd also never been as insulted by people I considered my friends—in fact, my only friends in a given place. I had contributed, making dinner and helping out. I had brought out a car that we could all use and made sure the burden didn't fall on his shoulders. Yet he only criticized me more forcefully than ever before. Roger was cruel and inconsiderate, I decided. Yet I couldn't call anyone to complain. I couldn't even mail a letter for another two or three weeks, nor would I get any mail for weeks. I felt completely cut off. I was used to dealing with personal problems by talking them through with someone. What was I going to do now? Complain to the Lewises?

That night I dreamt I was being murdered, stabbed with a metal blade through a chunk of my flesh.

The next day, I vowed I would arise anew and give the two of them more room. I wanted to make Roger realize his foolishness. In truth they were timid and afraid to say no—to stand up for their rights. But I refused to be the victim of Roger's "fed up" glances and hot, cruel temper. "The bastard, jerk," I thought, "hurting me in the name of consideration—as if he's been considerate." And to think I used to like and respect them.

The next night I didn't come back until after dark.

"I'm sorry," Roger said as I returned in the pitch black. "I didn't mean to chase you out of the house."

"I know it's difficult," I said.

"Let's just all be more considerate of one another. I know I am one of the worst offenders. I know my temper doesn't help the situation." We agreed to try, and he stepped outside. A few minutes later, I heard him yelling at Soba about something.

The following morning, when I talked to my assistants on the front porch, Roger was sulking again, hatred hanging over his face.

The next day, Koiya and Sana arrived, and Busakara and Sayuma quickly came into the house as well. I brought the meeting outside. Sayuma said he heard that Michael was leaving for America for six weeks, departing the beginning of next week. Sayuma then announced that he was going to stay in Goroka until the middle of next week.

"Why?"

"Worri bilong me" (from the English, "worry belong me," meaning "that's my concern").

"What is it?" I sensed something was up. Undoubtedly he feared what I was going to say to Michael about his performance. Michael probably knew anyway; Sayuma doesn't fool anyone. I reminded myself that I wouldn't have to work with him much longer. Still, I decided I'd better go to Goroka as well.

Sayuma went on ahead. I drove by myself.

I arrived at the Institute, feeling fearful at first about talking to Michael. Why was I here? Simply because of Sayuma? What were my questions? Still, I summoned my strength and went in to Michael's office to announce my presence. He was receptive, and said I'd be able to talk to him at 2 p.m.

I left to walk around town. When I returned and stood at Michael's door, Sayuma suddenly barged in and sat down plumply, with an official air, in the chair directly across from Michael's desk. He then proceeded to distort and lie about his kuru work, claiming that he had visited various villages. Michael seemed very pleased that Sayuma carried and used a notebook. But I wondered whether Michael realized that much of the accounting in it was false.

Sayuma also distorted the incident with the patient in Henganofi: "Bob pounded on the door and shouted open up, open up! The man opened the door, Bob went in to examine the patient. His brother came over very angry and told Bob to stop. The man had a knife, and I had to tell Bob not to continue." The moral: good Sayuma, stupid Bob. It was all a lie. I was amazed at the outrageousness of these stories, but ignored them. I chose to be tactful, assuming Michael would see through Sayuma's prevarications, and didn't tell Michael what really happened—as perhaps I should have. Next time I will, I told myself.

The Stone Age culture clearly developed the skill of psychological gaming into a tool as powerful as it is in the West—if not more so, given the few other resources that were available here. The local people's ceaseless requests for favors and gifts wore me down, leading me to accede in the end. The Fore were brilliant in their stratagems—in fact they were businessmen, and often little else. They occasionally sang songs and told stories—but the people here had no inquisitiveness into anything beyond their hamlet and personal material gain. The kids burst with warmth, in a refreshing contrast to the old men, who trod heavily with long, sour faces, played cards continually, didn't work, or speak to women, and constantly asked me for bags of rice, boots, and rides in my truck. The kids were alive to something beyond the hamlet fence, something other than material goods. How sad that they might end up like these older men. I sympathized with the men, but they tired me out, too.

In the car driving back from Goroka, I sensed that Sayuma was studying me. I shot back a glance at him. He quickly turned away and pretended to look out the window.

PART III   BEYOND
THE MOUNTAIN

Peaks

I decided to visit Mount Hagen to see if Senlado were part of a cluster. The trip would be long and potentially dangerous. I would have to cross through the Central Highlands, home of fierce warriors—the Chimbu—who regularly attacked and robbed cars that passed through. Recently, one Caucasian man had been killed in such an attack. On the other hand, it would be interesting to see this area. I was unsure how long it would take to travel there and find the patient—I could be gone for several weeks. Word that I was leaving spread quickly through the village, and in the late afternoon on the day before my departure, a group of men stopped by the house. "You i go?" Busakara asked.

"Yes but I'm coming back," I said. They relaxed, relieved. My presence here was important to them, despite all the difficulties I'd had with them. With their strong emotional connections to others, they would miss me.

"I have your raincoat marked," Busakara added, smiling. Through rainstorms and tropical squalls I had worn the poncho. Busakara somehow believed that because he wanted the poncho, it was his, as if the strength of his desire guaranteed his ownership. I sensed his strong attachment to it, and didn't want to hurt or disappoint him. He had been friendly and helpful, and never asked to get paid in return. It was the least I could do.

"Okay," I said.

"And I have your boots marked," Sayuma quickly added.

"We'll see," I replied.

I left early. I had arranged to take Sana with me, and had him meet me at the Purosa road *bung*. In Goroka, Wendy seemed depressed. I, too, felt cynical, heavy, and "unjoyful" around her. I didn't know if I was too dependent on her, asking her too many small, little humorless questions. Normally, she loved to laugh, yet lately was tired, and having trouble sleeping.

"I just read a great book," I mentioned. "*Wind, Sand, and Stars* by Saint-Exupery. It's beautifully written and very inspiring about living in a foreign country."

"Can I borrow it?" she asked.

"Of course." I realized that books here were special because the supply was limited. In the West, they must have once been cherished so as well, passed on from one person to the next.

My watch had broken in the bush. Wendy gave me the name of a small store owned by an Australian man, and the next day I went in. He sold a few cameras, looked at my watch, and said that if I left it with him for a few weeks, he would send it off to Australia to be fixed. I didn't have a choice, and gave it to him. (Months later, before I left New Guinea, when I stopped by the store to get my watch back, the owner still hadn't received it from Australia. I gave him my address in New York. A year afterward, it finally arrived for me back in the United States—still unfixed.)

Back at the Institute, I stopped by to say hello to Carol, the New Zealander whom I had spoken to previously.

"Do you want something?" she asked standing up. "I have work to do."

"Well I . . ."

"I thought you were going to call me when you arrived."

I should have said, "Well, I'm stopping by now." But I didn't want to be rude. "How are you doing?" I asked.

"I'm fed up with this country. The people can't even speak proper English. Everything's overpriced and difficult. No wonder the country's stuck in the fucking Stone Age! I'm ready to go home." She quickly turned and walked away.

I spoke to Kanaua about getting a car. "Sure," he said. "Take this one."

"It's in good shape?"

"Sure." Finally I left, glad to get away from Waisa, and now Goroka, too.

After several hours of driving toward Mount Hagen, I noticed three or four drops of white liquid on the window shield, but I didn't think much of them. Then another two drops flew up. I stopped the car. Steam hissed from under the hood. I got out to open it, and saw that water was boiling into the coolant container and then leaking out. To my horror, all of the white plastic fan blades had broken off. The water must have boiled after the fan ceased cooling the radiator. I was stuck in the middle of nowhere. A PMV stopped and the driver agreed to escort me back to a store. As I prepared to turn around, the front right tire and then the rear right one both shuddered into a ditch. Another truck had to come to pull me out. I was shaken, and furious Kanaua had permitted such an inadequate car to go on the road, and hadn't properly looked after the vehicle as he was paid to do.

I made it to a phone and called the Institute. "How could you have let a car like this go out?" I asked Ami.

"If the condition of the cars is such that they are able to drive on the road, then we let them out."

The police finally picked me up, and took me to a motor shop where wrecked trucks and scattered metal parts sprawled like ancient carcasses. A beer-bellied, but humble and friendly Australian waddled out. "She's a New Guinea car is she?" he said, slapping his hand against the metal, once white and now covered with rust and dried mud. "It'll take my boss, Rolf, and me a few days to fix."

I didn't know what I would do in the meantime. I phoned the local hospital. "I'm a doctor doing research on kuru and my car broke down," I told a man who answered. "I need a place to stay. Do you know if there are any hotels around?"

"None anywhere near here. You can stay over with my family if you want to, though. Just come by the hospital."

"Oh, thank you," I started to say, but he had already hung up.

It turned out that Dr. Simon Gifford, the man with whom I had spoken, had stayed with Michael in Goroka. His daughter had been born in Goroka as well. I realized I had made the mistake of introducing myself as a doctor over the phone, and quickly revealed the truth. Whether for that or another reason, however, he remained frigid and condescending—though he did display a dry English sense of humor.

His wife was nicer. "Make sure you go to Mount Wilhelm while you're here," she told me. "It's one of the few 15,000-foot mountains in the world you can hike up without having to wear a parka. And when else can you go to the tallest mountain in a country? On a clear day, you can see both the northern and southern coasts of the country. If you go, get Alex for a guide. He runs the one store in the village at the end of the road there. You can also go to the Bayir River Wildlife Sanctuary," she added. "Though it's sad, seeing birds of paradise trapped in cages. Still, nowhere else in the world can you see as many varieties."

When it was time to go I had some difficulty backing out of the driveway the huge truck that the garage had lent me as a replacement. It stalled several times.

"Don't run over the surgery staff," Simon said curtly.

"Thanks."

R olf, the head of the garage, repaired the car, though the fan remained quite battered. He was Belgian, in his 30s, short and thin, with a trim haircut and finely cut moustache. He'd come here to see the country nine years ago, but was now tired of it. "I can't wait to get out," he said. "But the money's good."

The drive was spectacular, especially the Dellum pass.

Mount Hagen was much less developed than Goroka. The Highway reached here only a few years before. (Indeed, the Highway was still being built, each year extended further and further west.) Here, natives wandered the streets wearing nothing but leaves around their waists. They didn't seem the least embarrassed or self-conscious. I had a Fore contact in Hagen named Paul, whom Sana knew. But when I found him, he told me he hadn't seen Senlado for a year. "I've heard, though," Paul said, "that he was at the Pentecostal mission outside of town."

I drove there, but they had never heard of him. I went back to town and checked the hospital files to see if he had visited with complaints, but the records weren't arranged alphabetically. I met a young doctor, Adam Wilson, but neither he nor the other physicians had seen any cases of kuru. I went to the gas station where Sana had heard that a man from Goroka, Kebo, was working, but he didn't know Senlado either.

I was frustrated, having come all this way. I headed to the center of town, to the market place. Women squatted on the ground before piles of bananas and betel nuts. Again, no luck. We drove by the one strip of stores, but didn't see Senlado or anyone else who might know him. Dispirited, I headed back to the car. I gave up hope and got in the car to

Markets in town: selling betel nuts.

Markets in town.

drive home. I had come all this way chasing just another false lead. It would be difficult to demonstrate that incubation periods could be identical.

Just then, Sana yelled something out. He had spotted a man from Kainantu.

The man had heard that Senlado was working at a coffee plantation outside of town. We went there. I lifted the gate and drove in, past rows and rows of lush trees. Nobody stopped us. The plantation sprawled for acres. Suddenly, Sana shouted to stop the car. A tall man was walking down a row. He was young and attractive, with a winning smile. "It's Senlado," Sana exclaimed. We went to talk to him.

"I was a part-time preacher here at first," he explained. "Associated with a mission. But I've been here at the plantation now for seven months."

"No one knew where you were," I said.

"That's because I never wrote to anyone back home." I asked him about symptoms. He and his wife had been sick in December and January—"skin heavy," tired, arms and legs pained, "as if something inside were eating me." But they hadn't had any tremors, or difficulties walking, moving, or eating. Then their symptoms resolved—without any medicine or other treatment. He looked fine now. He did not have kuru.

I drove back to town, dropping Sana off to catch a PMV back to Goroka. Then I stopped by the hospital, to ask where there was a hotel. I ran into Adam Wilson again, who told me there weren't any. But he invited me to stay over with him and his wife. Adam was a long-haired, young, liberal doctor—a refreshing change from the Giffords. He was also friends with Michael, and with the Towsons who had stayed at the Waisa. It was indeed a very small world in Papua New Guinea—among both nationals and expatriates, though the two worlds were different, and overlapped little. The isolation in this country made connections important—the Wilsons and I were *wontoks*.

The next morning, on the way out of town en route to Mount Wilhelm, the radiator started hissing rudely. I found a large leak that had been previously but ineffectively patched up. I turned around and went back to the hospital, where Adam suggested taking the car to Ella Motors.

I called Ami, who again used lots of bureaucratic language. "If the condition of the car is such. . ." At Ella Motors, they would accept a

Purchase Order Number only if we had an account. They were leaving anyway for lunch. I was told I would have to wait until they returned. I phoned Ami again. The Institute did have an account, it turned out, with Ella Motors in Goroka which was transferable.

The next day, I got back on the road. But that afternoon, the accelerator cable snapped. It had worn thin. The police again picked me up. I told them I'd ask the health department to tow me. I wanted to go to the District Health Officer, not Simon Gifford again. But the police, who barely spoke Pidgin, took me to the hospital and drove off. I entered and looked for phone to call Ami. He was sympathetic, but again not too helpful. As there was no hotel for hundreds of miles around, I had no choice but to call Simon Gifford.

I cringed as I dialed the number and explained what had happened. "Friday afternoon is a burden for everyone," he said. A District Health Officer truck came, and towed the car to Rolf for repair.

Simon drove with me to the garage in his own car, since he had to get it fixed as well.

"I think I'll just wait for it here," I announced when we arrived at the repair shop, "and just drive at night when it's done."

"Don't. Look, why don't you just stay over."

"I wouldn't want to put you out anymore."

"No, you wouldn't be putting us out. I hope I didn't give you that impression.'

"Oh no," I said, lying. "Not at all."

Back at the Giffords, we talked about Carol at the Institute, whom they knew and nicknamed Crabby, one minute spending time only with nationals, then, when they took advantage of her, cursing them out. I felt more comfortable with the Giffords this time. Their English reserve seemed partly self-protective in this difficult country.

In the morning I headed to Mount Wilhelm. I liked the idea of seeing the north and south coasts of Papua New Guinea at the same time from the summit. Before I left, I bought food and supplies at Steamships: pot, cup, spoon, toilet paper, soap, matches, tinned food, and a second flashlight (called a "torch" here) for my guide to use. The road, twisting narrowly around the edges of canyons, was the most dangerous one I had been on yet. I concentrated fiercely, my palms sweating, my eyes scanning the stretch of dirt immediately in front of me and on the sides. I drove through steep gorges and past rock outcroppings, eroded and weathered, sculpted over thousands of years by rain. Scattered between the rocks, the Chimbu built houses on vertical supports. Plants sprouted from the roofs.

Several hours later, I arrived at the village. I pulled up and asked for Alex. Simon had advised me not to let a guide charge me more than about 12 kina. But Alex now wanted me to pay 15 kina for hiking up and 9 for returning down, totaling 24 kina in all—a steep price. I offered 3 instead of 9 for coming down, and he agreed. I also suggested selling him my torch for 1 kina, cup and spoon for 1 kina, and pot for 2 kina. Again he agreed. Thus I would pay him 11 kina for the hike up and 3 kina for the return, for a total of 14 kina. I could probably have bid him down more, since he wouldn't carry my sleeping bag, finding it too bulky.

We started our ascent. The tropical rain forest resembled others I had seen, but higher, above the cloud line, a vast airy space opened around us. In a long valley between two soaring ridges, we strolled through a corridor of grasses, mosses, and swamps. At one end of the valley, a wide waterfall sheered down, slicing through a wall of stone. The fresh white water surged down the center of the valley, into a wide gap of clouds at the far end. Lesser peaks hid far below in the distance.

Alex was kind and helpful, but there was not much to talk to him about—nor did he expect me to say much. The climb was exhausting but exhilarating. We arrived at an A-frame shelter, where we would spend the night. The shelter was built of two tilted wooden sides meeting together at the top—a wooden tent. It contained mattresses and was cozy. In front, here at 12,000 feet spread a clear, cold lake with steep mountains rising all around. The tranquil waters, untouched by man, mysteriously mirrored the surrounding landscape. I felt I had intruded on a hidden paradise. I boiled water and made hot tea with sugar. I opened some canned meat and vegetable stew from Denmark—a delicacy I bought at Steamships to treat myself.

At 2:15 a.m., Alex woke me. We set out in the dark, following the small round spot of light from my flashlight. With each step I took, my boot sunk quietly into the muddy ground. Several hours later, the sun began to emerge. Fields unfurled all around us in this virgin land—a land before time. Here was another extreme, another end of the island— the top. Below us, thick white clouds parted—solid masses ringed by

Mt. Wilhelm.

The author on Mt. Wilhelm.

mountains. A white lake of clouds nestled in this hilly landscape, as if solid white water. As the sky slowly brightened, yellow streaks ran delicately overhead. A few hours later, higher up, soft blue phosphorescent oceans of mist surrounded us. Across the sloping landscape, fragile but sturdy bush scuttled. Amid tiny leaves on the ground, baby buds bloomed—pinches of yellow, white, pink, and purple against the grey stone.

We passed a plane wreck from World War II—scattered metal along the side of a mountain. I wondered all the more how odd it must have been for Stone Agers to see—even from a distance—such birds in flames, falling out of the sky in torrents of smoke. The smaller pieces of the wreckage had been removed.

I was surprised to see a bronze plaque on the side of a rock, informing us that here were found the last possessions of a patrol man who had disappeared in the 1960s while exploring these peaks. Below the plaque a box remained—the last sign of his presence on earth—a long rectangular case of brown leather with a leather handle, used for carrying a piece of surveying equipment. I knelt down to open the case—surprised it was still here. The wind gushed past my ears as I crouched on the bare ground. I unlocked the box. Worn silk—once white, now yellowed to a pale cream—layered the interior. The box was empty. I stood up and stretched my back. For miles and miles around lay nothing but desolation. I looked beyond the peak and realized he could have gone anywhere in this inhospitable land. Not even his remains had ever been found—so deserted was this region of the world.

I continued walking, but every few yards had to stop, short of breath. Afterward, I realized that it was because the air was thinner up there. Cold wind soon whipped around me. Across the valleys beneath me, clouds now raced, faster than those further below on Earth. It began to get even colder. Higher up, only cracks of color sliced through the endless block of crystal sky above. I had trouble standing because of the biting wind, the fatigue, and the low oxygen. The trip ascending this mountain resembled a journey north from the equator to the arctic. We were also traveling backward through time and evolution—from human beings to animals, to forest, to scrubby grasses, to fine, velvety green mosses, to small spots of grey and white lichen (all that would grow), to barren rock.

Hours later we arrived at the summit—a desolate, inhospitable place. Bare bones of igneous rock barred any life. I stayed at the top as long as I could. Silence radiated through these lifeless tracks. Continental drift was pushing these mountains ever higher. This landscape resembled Mars or Venus more than it did Earth.

For the first time, I felt the Earth to be a planet.

By the time we returned to the base village, it was late, and too dark to drive. The road was dangerous, and "rascals," thieving gangs of young boys, were said to prowl it at night. I planned to sleep in the truck that night. Still, I was glad to be among other people again. I gave my guide the amount of money we had agreed on—14 kina, along with the pots, pans, silverware, flashlight, and compass I had bought especially for the trip. He looked down at the roll of colored paper in his hand, but didn't know how to count money. He smiled and gave it to a friend of his, who quickly counted it. "It's not enough," his friend said.

"But it's what we agreed on," I protested

"No. It should be 24 kina."

"No," I said, "We had agreed to this amount since he would be getting the equipment."

"No, he needs 24 kina," the friend insisted.

"But that's not what we agreed to."

"No, you need to pay more!"

The friend didn't understand that I could have somehow renegotiated the terms. The concept of bargaining didn't exist to him. Reluctant to pay, but afraid because I'd be sleeping in the truck parked in their village during the night, I gave in. Afterward, when I told Wendy how much I shelled out, she said, "That's the most I've ever heard of. No one will ever be able to afford to hike up Wilhelm again!"

On the way back to Goroka the next morning, the car broke down for the fourth time. This breakdown would take several days to fix, so I decided to fly back to Goroka instead. I'd never felt as ready to board a plane.

The tiny local airstrip was positioned at the edge of a cliff. I checked in, and was assigned the copilot's seat in the front of the six-passenger propeller aircraft. The plane wheeled forward straight ahead along the ground. Then, suddenly, the ground below us fell away and disappeared. As a result, having left the cliff, we were in the middle of the air, but the plane was dropping. In terror, I looked out the window as we plummeted down into the valley. Before me the controls and altimeters swung back and forth, tottering. The steering wheel rose, turned, and fell. But the cliff was high, and the valley bottom still far below. Slowly, we gained altitude. This cliff top was the largest flat surface for hundreds of miles around. In the days before the road was built, the airstrip was the only way to reach this region. Only later did I hear that several planes had crashed here.

Coasts

"You look like you could use some *malolo*," Michael told me when I arrived back in Goroka. "Everyone who comes here to PNG does after a while." Months of trekking, traveling, and bartering had taken their toll on me. Leeches, snakes, unclean food and water, hazardous driving, and trying conditions weighed increasingly on my mind, heightening my vigilance at all times. A few weeks earlier, two expats in the Highlands had suddenly died from the potentially toxic antimalarial medications we were all taking. The tropics also did strange things to people, psychologically—"tropical rot" could set in from being isolated and besieged for months. When the rare airplane roared across the Highland skies, I'd gaze up, longing to escape these landlocked valleys.

"Why not go to the coast for awhile?" Michael said now. "Put your feet up and reflect on what you've been finding. The Institute has a compound where you can stay. Maybe you could write a paper on your research."

"Okay," I said. "How would I get there?" I was still nervous about travel.

"Probably fly," said Michael. The drive is beautiful, but during the rainy season the dirt road can get washed out and take weeks. It can also be dangerous."

A flight left for the coast every few days and I made a reservation. I decided to stay there for a fortnight before returning to finish my field work. On the plane I felt nervous at first, but as we ascended my spirits lifted. For a while we flew in a complete void, grey covering all the windows, but as we climbed further, the clouds broke and the Goroka Valley soon lay far below us.

Behind me, an older man sat barefoot, balancing a huge bushel of yams on his lap. The bag loomed up over his head, his arms reaching about the bag's circumference, but not making it all the way around. "They're for my son on the coast," he told me in Pidgin. "He works in a coffee plantation and sent me money to come visit him and bring him *kaukau*. The *kaukau* on the coast are nowhere as good as up here in the Highlands." The man had never been on an airplane before. He sat looking straight before him at his bushel, never daring to glance out the window.

The aircraft slipped through the one northern pass in the mountains, which I knew from maps to be the Bena Bena gap. Mountains stared at us in the face from their heights. Beyond the pass, the land suddenly swept down. I was able to spot the thin road lacing back and forth. The Ramu River also rolled down from the mountains, curling around leisurely—as if a loose gold thread, dropped from the sky.

Finally, in the distance, the endless blue ocean spread, calming and refreshing. As I had grown up in New York, I was used to being near the sea. Mountain ranges now rambled and petered out before the vast Pacific. To one side, I spotted Galapassi Point, which I recognized from a book about the first anthropologist in New Guinea—a Russian nobleman named Mikloucho-Maclay—who in the late nineteenth century, had ventured there, dropped off by a Russian trawler that returned a few years later to pick him up. He could not penetrate any further into the country, and believed the interior to be uninhabited.

Suddenly, straight ahead in the distance, rose a strange sight—a single blue peak. Above the otherwise calm surface of hills and wispy clouds soared a giant mound—steep, symmetrical, and solitary amidst the distant sea and haze.

"What's that?" I asked the pilot over the roar of the propeller engine.

"Karkar Island. A holy place. The local people worship the mountain. It is a volcano." It now spread across a wide girth of the Pacific, the two sides perfect mirror images—their geometric slopes ascending slowly at first, then more steeply. I noticed how much volcanoes stand out as the only large, symmetrical shapes in nature. As we flew closer, the mountain increasingly dominated the flat horizon. Behind us, Mount Wilhelm and the Bismarck Range seemed small and craggy by comparison, mingled with clouds. The pilot told me that Karkar had exploded six years ago, and again two years ago. The island now stood mysteriously over the landscape—a portender of doom in this new, unstable world.

We finally descended into the town of Madang. The small port, first settled by the Germans before World War I, lay at the end of a short peninsula that seemed to push itself out into the Pacific. From the air, the town looked merely like tiny bits of tinsel scattered at the side of the sea, across lush green inlets, bays, islands, and lagoons. All around spread the endless ocean, a huge vastness mirroring the sky. In the distance, several miles offshore, Karkar hovered.

We landed at one end of the town on the single-lane dirt air strip, which ran parallel to the water, separated only by a thin ribbon of sand washed by the waves. I telephoned for a cab. It arrived an hour later, and took me to the Institute of Medical Research compound outside of town. There I met several other young researchers—mostly Australian and English—and was assigned a small bedroom. I would be sharing a kitchen and living room with three Australians. Brian and his girlfriend, Frances, both tall and blond, were conducting research on malaria. Sarah, shorter, with long, straight brown hair and dark brown eyes, worked as a physical therapist at the local hospital.

I was also given a small office. I soon unpacked my papers and sat down to try to work, separating out all my notes and trying to put together genealogies of the patients I had seen over the months. But I was hot and sweaty, and drained from the climate and the trip.

Sarah stopped by. "Want to go to lunch and to the beach with us?" she asked.

"Sounds great."

Brian, Frances, and Nick Sullivan, a heavy-set Englishman, joined us. I felt like I was back in college, surrounded by friends my age. I hadn't engaged in any such social activities for months.

"Where should we go?" Sarah asked.

"How about the Madang Club?" Nick said.

We drove to an old colonial building, constructed before World War II at the edge of a lagoon. We entered through a front portico of white wooden columns. Large overhead fans whirred around the cool, dark interior. Light spilled in through windows covered with mosquito mesh and shaded by shutters. The crowd, all White with white hair, looked left over from colonial days. A sign over the bar announced that women were permitted here only with men who were members.

Our small group slipped out the back onto an otherwise empty terrace, and gathered lounge chairs together to form a small circle. Around the terrace, banana and palm trees swayed lightly in the breeze, and luminous orange leaves burst out from bushes. We ordered hamburgers and chips and sat drinking cold beer. "This," Nick said, his bottle in hand, looking out over the lagoon and distant blue mountains, "is what makes the tropics all worthwhile."

"So what made *you* come to New Guinea?" Sarah turned to me and asked.

"I wanted to go somewhere far away and try doing research. How about yourself?"

"I've always wanted to come here—since I was a little girl growing up in Australia. It's always had some special appeal to me. I don't know why. Melbourne's pretty boring, and I was sick of home. I figure once I go back to Australia, I'll end up staying for many years. I went to India two years ago and then decided to come here. My two brothers also love traveling. They just visited Nepal."

I envied their adventures, but then realized that I, too, was now exploring the world. "How's your research been going?" Sarah asked abruptly.

"Okay," I said. "But I got tired, which is partly why I came here—I needed a break." I told her about my work, the clusters of patients I had found, and the elderly women who had survived the epidemic.

"That's amazing," she said. "You should go back and find more of the older women and study them. Perhaps draw blood samples from them. You may find something important. Besides, this is a very unique time here. The traditional world is quickly disappearing. The government is corrupt, and has no interest in preserving anything of value. Do you know the government sells strip-mining rights to foreign companies at the drop of a hat? Huge amounts of money go to payoffs. The rain forest will one day all disappear. You'll look back at this year very glad you have been here and done this."

I was surprised how much I had lost sight of that.

"What's *your* work been like?" I asked her.

"The hospital here has never had a physical therapist before. So they're very grateful for whatever I'm able to do. There's a lot of work though, since I'm the only one. I'm teaching one man to walk, who couldn't before. I also play wheelchair basketball with a patient who just used to sit on the ward all day."

After eating, we strolled around the club grounds. I ended up chatting with Sarah further, as I hadn't to anyone for months, about movies (we both liked Woody Allen), books, traveling and our families. I realized how much I had missed such conversations, especially after talking in Pidgin for days on end. I felt I was returning to my former self. "Where in the States are you from?" she asked.

"New York."

She leaned closer and lowered her voice. "Are you Jewish?"

Her question surprised me. "Yes," I answered.

"I am too. Do you realize we're probably the only two Jews in all of New Guinea? I haven't met any others." I hadn't either. Moreover, being Jewish had never been much a part of my identity back home, and I had never been as aware of being seen as different.

After lunch, we all went to the beach. The five of us crowded into Brian's small car. Frances, in a bikini, Sarah, in a black one-piece bathing suit, and I squeezed into the back. Brian drove us down a dirt road to a hidden cove—a calm blue lagoon bordered by an arc of palm trees and bright fluorescent yellow shrubs sprouting up from white sands. No one else was around. I had never seen a beach as unmarred by people or hotels. "My God," I exclaimed. "It's just like Gilligan's Island"—my only point of reference. They all laughed, and we leapt into the water, except for Sarah, who sat on the beach watching.

"Aren't you coming in?" I asked her.

"I'm not very good at swimming."

"It's not very deep."

"I'm still nervous."

"Don't be. It's not that bad." I liked to swim—though I hadn't been able to since arriving in PNG—and had completed Red Cross training as a lifeguard several years earlier. At last, Sarah walked over and waded in.

I tried teaching her how to swim better, showing her how to synchronize her breathing. Then we put on snorkeling masks and flippers, and dove down into a world of brightly colored fish. I pointed to angel fish—black, with thin, wavy white and yellow stripes—sweeping by, trailing long, white tails gracefully behind them. Lime-green fish dotted

The coast.

Views from Madang.

The author in Madang.

Madang at sunset.

with white spots glided by in families—one parent in front, the other bringing up the rear. Zebra fish poked at fingers of live pink coral. Squid with long, pointy noses looked like they were swimming backward. Schools of green and purple fish flashed silver when they turned. Plants waved in the current. Sarah pointed at a violet starfish sprawling on the rocks as if sunning, glowing as if painted on. Stalks and spiraling bowls of rose-colored coral covered crevices and caverns. Fish fluttered like butterflies about these flowers of the deep. We were traveling through a primitive kingdom, preserved from eons ago.

Finally we headed back to shore, and sunbathed. Frances had brought along a snack for us of coconuts and oranges, neither of which I had had in weeks. We lay on the beach all afternoon. Gradually the sun set amidst deep reds, mauves, and golds—the colors spilling on and over themselves in the gentle lagoon like honey pouring out of a jar. From the trees around us, birds chirped and sang in the growing still-ness. I relaxed as I hadn't in months. I realized how much the tropics embodied dreams—unchoked by the cold grey rain and snow of north-ern climes.

Back at the Institute, I washed and dried my laundry in machines. The batch of freshly cleaned clothes, as I carried them back to my room from the electric dryer, warmed my arms.

In the early evening I was sitting, writing in my journal when Sarah wandered by.

"What are you doing? she asked, poking her head through my door.

"Writing about my time here," I said, somewhat embarrassed.

"Do you want to be a writer?"

"I guess so." I had never dared to talk about it much, or let myself consider it too far.

"I'd be interested in hearing something you've written."

I read her aloud some passages about my experiences in the High-lands. I had never read my work to anyone before, and enjoyed having an eager listener.

"That's beautiful," she said. "You should keep writing about all of this. Write a book about it one day."

"You think?" I suppose the idea lurked somewhere in the back of my mind, but I had never articulated it or considered it possible.

"Sure." She smiled. "Why not?"

The next day, at 11 a.m., Sarah came by again and asked if I wanted to take a tea break. We set up two wicker chairs and a small table outside under the shade of a huge banana tree, and sipped a pot of freshly made tea with lemon. The countryside rolled around us.

"I just got letters from my mother and my ex-boyfriend back in Australia," she said. "My mother writes about all the new clothes she's just bought. My old boyfriend is an OB-GYN resident. I think he's basically in it for the money. In the end, I simply couldn't go out with someone like that. Most doctors just have blinders on."

"That's why I've had doubts about it," I said. "It's also a long haul."

"But what could be more fulfilling? I think it would be a great education, and you can do many things with it afterwards."

"I've also thought about more humanistic pursuits."

"But you can do those, too."

That night, the Institute researchers were performing a play, *Wait Until Dark*. Since Madang had no televisions, movies, or video rental stores, these expats put on plays to entertain themselves. Brian and Sarah helped with the sets and lighting. Nick and Frances acted. Afterward, we all went out for drinks and got drunk. It was fun to be again with bright people of age and background similar to my own.

I drove back to the Institute with Brian. On the road into town, however, four policemen stopped us, checking cars. They barked orders in Pidgin to put on our wipers, and turn on the lights and directional signals. One of the front headlights didn't work.

"Em inap?" Brian asked—Pidgin for, "Is that enough?"

"Bai mi kisim yu long offis bilong polis na suspend registration bilong car," one of the policemen snapped threateningly ("I will take you to the police station and suspend the car's registration!")

We had to pull over and wait. Another car, whose signals didn't work, but which was driven by a New Guinean, was allowed to pass. Brian went up to the officer and asked if we could fix the light and return in an hour. Reluctantly, he agreed. We headed off and got it fixed. But when we came back an hour later, the cops were gone.

Over the next week, encouraged by Sarah, I worked on a draft of a paper based on my work. I realized I had gathered more and better data than I had thought. I had discovered three clusters of patients with identical incubation periods. No one had ever known if such clusters actually occurred. It was exciting to uncover a bit of nature itself. God may not exist, but if he does, the surest evidence is in Nature—plants, animals, and other life forms, growing, taking shape, and multiplying, following their own hidden laws.

Although I had thought a lot about my research, writing a paper focused my thoughts, showing me weaknesses in my analysis and questions that still needed to be answered. I saw now that I was missing key bits of information that would be important to go back to collect. I discovered that two recent patients, Mabamo and Paul, listed on separate genealogies, were in fact cousins. It appeared that they had each attended only two feasts—the same two—in their lifetimes, held one year apart, along with a relative, Timani, who was born a few days after Paul. Timani was Paul's *nagaiya*, which in *tok ples* means "my umbilical cord," and refers to children born within a few days of each other and kept with their mothers in the birth hut until their umbilical cords disappear. Timani became sick and died six years after the feast. I realized I could go back and collect further details about these feasts.

While in Madang, I also read much of the first volume of Proust, which I had borrowed from Michael Alpers. I discussed the book with Sarah, who had finished it not long ago. "It helped me to appreciate much more the beauty around here," she said. "It's one of those books that makes you see things differently." I agreed: the novel's intense descriptions of colors, sights, and smells inspired me to look at this strange land more clearly. Frances listened and said she wanted to borrow the book as soon as I was done.

The following week, Easter arrived. Everyone was going on a trip to Karkar, to hike up to the top of the volcano. The natives never hiked up it, because the mountain was a place of *maselai*.

"Maybe we shouldn't go," Sarah said, as we all sat around one evening.

"Nonsense," Nick answered. "That's just superstition. Hogwash. I think it would be fun. Besides, when else are you going to get to see a live volcano?"

"But it would bother the natives," Sarah said.

"So?"

This conversation made me uneasy.

"Maybe we can meet some of the natives while we're there," Sarah suggested, trying to find some consolation. Nick looked dubious. "Anyway, isn't it dangerous?" she added.

"It's just a mountain," Nick said. "We're bright; we can handle it."

"Why not"? Brian added.

"Maybe I'll join you all," I said to Sarah.

"I wouldn't if I were you," she said, "I'd go to the Sepik River instead. Especially if you don't think you'll be back to the New Guinea coast again anytime soon. Artistically, it's the most interesting place in the South Pacific—the home of the best carvers. I don't think Karkar will be nearly as worthwhile. But I've never been there, and it's important to explore new places. After all, you only live once."

Encouraged by her, I opted for the Sepik. This region had been in contact with the West since the late nineteenth century. I wondered how it managed to integrate traditional and modern cultures. How had tribal beliefs and views changed? The Highlands were only just now beginning to face these issues.

A few days later I flew out of Madang. My time there felt short, but immensely enjoyable. For the first time in this country, I had made friends. Below the belly of the plane, the Ramu River snaked leisurely to the coast, curling around like the heavy outline of a bird. The landscape, mottled green and brown, resembled army camouflage. Jumbled lakes and oddly shaped patches glistened brown and forest green. It was hard to tell what was water and what was land. The Highlands hid behind the clouds, far away.

Floating Islands

The airstrip where our plane set down stretched between gentle hills and forests of palm trees. The road to the airport ran along the beach, separated by a narrow strip of land fifty yards across, covered with sprouting gardens of beans and lawns of clover and wild grass. Around the airstrip, large families of nationals lived in corrugated iron shacks. Through the sides of the houses, large holes rusted. On the roofs, metal sheets jutted out off the eaves at various angles and flapped across one another in the wind. Beside one house, seven-foot-high wooden carvings of human figures with large, flat heads and skeletal bodies stared blankly. At the side of the hut, a man hacked away, carving another figure.

The area of the Sepik River near the coast had had more contact with Westerners than villages upstream. The further upstream the village was, the more isolated the people. I decided to travel to the furthest point accessible by road, which was midriver. I would subsequently journey by canoe as far upriver as I could, then turn around and return by water toward the mouth. On the main road near the airport, I waited several hours for a PMV, but none drove by. I then tried hitchhiking. Every time a car approached I leapt up and stuck my thumb out, but the drivers, concentrating their eyes on the dirt road, waved back as if I had merely said hello. I suddenly realized that they didn't know what hitchhiking was. It simply didn't exist here. In this country, with PMVs and few private cars, people paid for rides. Everyone would hitchhike if they could.

PMVs (private motor vehicles).

At last, a PMV came rattling around the bend. I stood out in the middle of the road to flag it down. It was a crowded flatbed truck. Three people sat in the front; twenty-five others squat in back. The driver tried to pass me without stopping. But I was desperate, and stayed out in the middle of the road, blocking his path. He finally braked and agreed reluctantly to take me along.

After we had been driving for twenty minutes on the bumpy road, rain began to slash down on the open truck. Some other passengers found a large plastic tarp that we pulled over ourselves, arms reaching all around. We had fun laughing and joking, and were soon huddled inside the tent.

The trip took five hours. We passed several large *haus tambaran*, traditional houses of the spirits, used for ceremonies. When we reached the first town, I found a room at a tiny motel without electricity. Blackouts frequently occurred here and lasted for days. For dinner I had to buy food from a *stua* down the road—cheese, crackers, a hard-boiled egg, *kaukau*, and peanuts. Outside, rain continued to beat down. The hotel was dark and decaying; the walls swayed in the wind.

In the morning I left the hotel and walked around the town. The traditional houses leaned forward here. Curious, I crawled inside one. Light streaked through small holes in the walls, filtering through a dark, misty space filled with mysterious carved figures.

Back outside, I stood in the road for hours waiting for a vehicle heading toward the river, but none came by. Finally, a truck approached. Again, I stood out in the middle of the road to block it, and begged the driver to give me a lift. He did, but was going only part of the way. Soon I was stranded once more, in the middle of a vast breezy plain rimmed by blue mountains. Scattered lone clumps of odd trees twisted toward the vast sky.

At last, another truck rumbled by. The driver would take me, he said, but he had two *wontoks* in the cabin already. I would have to stand on the running boards and hold on. Luckily, I had already done this with Mr. Lewis, and so agreed. But the truck, now on the flat road, sped along at forty-five miles per hour. I gripped the metal ridge on top of the door as fiercely as I could as the landscape sailed by. Clear, cool air swept the land. In retrospect, I realized the wind came from the river, not far away.

The driver stopped at a riverside town at the end of the road and I had to get out. A handful of stores stood in a row, each fifty yards apart. At the end of the road, the Sepik River stretched. Its width surprised me. It was miles across, and flowed slowly with a spirit and pace of its own, leisurely but surely, defying the land. Yet I wouldn't have known it was there until I stood before it. It is, in fact, one of great rivers of the world—yet had not even been discovered by the outside world until the 1880s. I have since seen many great waterways—the Nile, the Rhine, the Mississippi, the Ganges and the Yangtze. But here I experienced for the first time a certain feeling of awe. It seemed to have a life, to be a beast of its own. This river connected to all other waters on the planet—to the South Pacific, the Indian Ocean, eventually the Atlantic.

At the road's end, moored to a tree, floated two canoes with outboard motors, rumored to be going to Ambunti—my destination. "Can I get a lift with you?" I asked a man sitting by an oar in one of them.

"Well, we're not ready to go yet," one of the "bossmen" or owners told me. "It depends on how much space we have."

I heard there was a crocodile farm not far away, and I went to find a phone to call and see if I could visit. Suddenly I heard a motor from the river. The boat had left, but the other vessel was still there. I went back and spoke to a short man on board, who said the bossman wasn't present. I waited. When he arrived, I asked if he'd take me. There was no hotel in this town. The nearest one was where I had stayed the previous night. If I didn't get on this boat, I might have to return there. But the boatman turned me down, claiming he had too much cargo. Then he told me he wasn't going all the way to Ambunti. I suspected a lie, and said I'd go wherever he was going. He then said he'd have to ask his bossman, though I was sure *he* was the bossman. He left and came back, saying the bossman said no. I replied that I'd ask the bossman myself. I assumed the "bossman" was probably just another passenger, but I wanted to clarify the situation.

I began to walk over to a tree stump to sit and wait, when the boat quickly started its motor. Two passengers jumped on board and the boat rapidly took off. Crestfallen, I cursed them. Finally, a student came up to me, and said his brother might be coming to get him and could give me a lift. At this point, I would have been quite surprised if he actually came. I sat by the edge of the river and waited. Mosquitoes found me, and I swatted them as they bit my arms and legs. This was a malaria-infested region—with Fansidar resistance. Luckily, I had been doubling my dose since arriving in Madang.

A drunken man strolled over. "What are you doing?"

"Waiting for a boat to Ambunti."

"What are you doing here in New Guinea?"

"Medical research in the Highlands."

"Are you a doctor?"

"A medical student."

"I don't believe you."

"I am."

"I know of a boat going tonight to Ambunti."

"Really? Great."

"It will be expensive, though. You'd better have plenty of money."

"I do."

"It's best to carry traveler's checks there, you know. It's also dangerous here." He pulled out a knife. "Pigs and people get cut and are never seen again." He may have just been scheming, but I was frightened. I rose, wary. Just then another boat pulled up, and I scurried toward it.

"Are you going to Ambunti?" I asked quickly and in desperation.

"Yes," the man in the boat said, nonplussed.

"Can you give me a ride?"

"Yes," he said, not at all bothered by my request.

I jumped in, relieved. I crouched down in the canoe—a long dugout tree with a motor in the rear—as we sped out into the middle of the river. Just then, the motor puttered out. We paddled back to shore by hand, and I stayed in the canoe with another passenger while the owner fixed the engine. At last we set out once more and headed upriver. Again we glided past the land on either side. Suddenly, the motor gargled and broke down, this time a little further than before, and we had to return to shore. The owner pulled up the motor and tinkered with it for an hour. No other canoes arrived or left. I had no choice now but to remain with this boat.

Finally, we got away. The sun was now setting, and the leaves on the shore blazed with bright colors fanning across clouds and deserted azure mountains. Deep purple silhouettes of tall grasses and shadows of trees reflected in the still waters, as the sky turned mauve. Behind the scalloped shapes of trees on the shore, clouds shone mysteriously with white light. A thin band of tall grasses and low clumps of trees separated the gaping voids of sky and water. Soon a star reflected in the water, shining equally in both water and sky. Moonlight began to illuminate the land. On the bottom of the long dugout canoe, from front to back, a smooth thin line of water stretched, licked by the white moon. The concave shell of the boat floated, its bottom below the clear surface of the river. A breeze blew gently over my tired body and through my sweaty hair. As the moon rose, my skin slowly cooled. I'd never felt such a reversal of fortune as I experienced now leaving the shore and its frustrations behind.

Strange noises rose from the banks—birds chanted and screeched, and frogs croaked in shrill, unfamiliar tones. Cicada trills echoed over the waters as if calling to someone. Startled hens rose up and darted back down toward the ground, blaring and casting shadows in the moonlight. High grasses and silhouettes of strange trees crouched in the darkness like stranded dinosaurs. We glided quietly past tiny silent hamlets. This vast river meandered infinitely, seemingly lost.

At midnight we reached Ambunti. I was glad to arrive, and soon found a school that rented out a room for the rare overnight visitor. In the morning the village was quiet but beautiful at the side of the empty river, surrounded by forest and ringed by mountains. I visited the *haus tambaran*. At either end, masks stared down under the V-shaped roof. Bulbous eyes sprung from heads. The nostrils were pierced—as were most men's—yet here made handles. Smoke floated through the dark interior. Men sat, unhurried, indifferent, stoned on betel nut. In the center lay drums carved from hollowed-out tree trunks. Heads decorated the handles protruding from either end. Wide eyes stared with awe. Amidst the clutter of skillful but lifeless art, these faces stood out. At one end, two painted straw masks hung from the ceiling in a shower of colorful baskets and straw. The masks had looped straw noses, as if pierced. They were *tumbuna*—sacred relics—and could not be photographed. The God here is a woman, and worship is an exclusively male affair. These houses were usually *haus boi*—where boys, once initiated, lived.

Back on the town's few dirt streets I met a missionary: "Seen one *haus tambaran*, seen them all," he said. "The natives here have a 'coastal attitude'; they don't care. There's gold and copper in the mountains behind us, but the natives are too lazy to get it. Christian

missionaries were here and left. I'm the last one remaining. The people are now returning to spirit worship." Christianity had failed in its
first attempt here; in other words, the people had eventually rejected
its teachings and thrown the church out. "*Haus tambaran* are now
used once again."

"That doesn't surprise me," I said. The missionary looked at me,
shocked.

When I walked away, villagers approached me. "Do you know
him? Is he a *wontok* of yours?"

"No."

"Good," they sighed. "He is an evil man." Most Westerners in this
region were connected with a church. I realized that some of the difficulty I had been having in this region might have resulted from the as-
sumption that I, too, was a missionary.

By the shore, a woman sat weaving a funnel-shaped fishing basket,
cleverly designed to allow fish to enter but not leave. Her husband
showed me crayfish and small five-inch silver fish he had caught using
such a device. Other women wove *bilums* of light brown native bush
rope and dyed purple strands

The next morning, a native was canoeing down river and I joined
him. Overhead, light grey herons, white imperial pigeons, and willy wag
tails darted. Swallows flitted nervously about. White egrets with feathery
wings swept through the air, their yellow feet floating behind them.
Blackbirds sailed on curved wings. Snake birds flapped, and dove with
long, thin beaks into the water, spearing fish. Dollar birds watched,
perched on trees. Whistling kites stood long-legged in the reeds.

We canoed to Pelembi—a town once located on a bend in the river,
until the Sepik had changed course, leaving the town on a connecting
side lake. Houses perched over the waters that mirrored the purple sky.
Between sky and water shimmered a delicate thin line of luminous, almost florescent green grass. On either side of lush green carpets, ponds
lay like sheets of reflecting glass. Lilies leisurely sprawled.

I crossed a bridge of poles supporting angled planks. A long green
lawn opened up. At one end stood a *haus tambaran.* I climbed up the
stairs—between the legs of a woman, her back a crocodile and her but-
tocks a pig's head. To enter the house, the visitor had to enter this
woman's body. The house embodied the initiation rites that turned
boys into men.

Outside, in the middle of the lawn an old stone leaned crookedly—a stone *bilong tumbuna*, larger than any stone found near here. The people revered even bits of an older culture—assumed not to be man-made, because reflective of a different society. Further down the green rose a line of tall poles richly carved on top with faces and decorative patterns. Two houses tilted, collapsing, toward the earth. This older village, left behind by the river, preserved the foundations of its past more than any other I had seen. I bought a carved story board—human figures turning into crocodiles and birds. A *maselai* snake in a tree surrounded unseeing women in the forest while the big man of the village stood by with his wife. A *haus tambaran* sat in the background. I also bought a standing sculpture of Yanegi—a man of a legend who knew magic. He killed his brother-in-law, whose line was from elsewhere. Yanegi's body then split down the middle. He lost one leg, and half his penis and tongue, i.e., his sexual potency, power of speech, and walking—presumably the three most important of man's abilities—the capacity to walk, talk, and fuck. The sculpture showed half of Yanegi's ribs, one leg, and the side of his genitalia—a human figure simultaneously in cross section, elevation, and profile. The carving combined these three projections (which are used in CT scans of the human body as well). In the West, it wasn't until Leonardo daVinci that Europeans began to experiment with representing the body in each of these three planes.

I proceeded further down the river by canoe. In the next village, carvings *bilong tumbuna* lay in the *haus tambaran* like Catholic relics. In the dusty upper floor of the *haus tambaran*, a man pointed to a costume of dried grass, leaves, and gourds hanging like an old tattered wedding dress. "Em bilong tumbuna tru, em oldpela tru" ("It is truly traditional, very old"), he said about what seemed to me to be a pile of dried vegetation. Yet to him it was imbued with magic, glowing in the thin light that filtered through the heavy smoke smoldering from the ground. Sacred art does not have to be good art—merely old or viewed as magical.

I asked what other art they had. The man brought out carvings of stylized faces with long noses bridging down to chins, as if the noses were pierced. The long, snakelike noses also resembled bird's beaks. The villagers then brought out even older pieces—feathered, fragile, and a few yards high. These would be difficult to transport out to the coast by canoe. I could try, but imagined I would instead be able to buy similar pieces closer to the coast.

Carved faces and crocodiles covered the poles supporting the *haus tambaran*. The villagers would have sold parts of the *haus tambaran* itself—for the right price. Imagine walking in and buying the stained glass windows at Chartres. I couldn't believe they sold me heads *bilong tumbuna*. But if they hadn't, I thought, these relics would have just rotted here, or been bought by someone who might appreciate them less.

At the next village, I paid one kina to see two beautiful masks *bilong tumbuna*, carefully wrapped in broad dried banana leaves. The sculptures had long downward-pointing noses, encrusted with small magenta and white shells. The small masks captured an aspect of the typical New Guinean face—the curvature of nose and prominent eyebrows.

Back on the river, the sunset was the most spectacular I'd ever seen. Gold burst forth from deep red clouds, amber in the sunlight—all mirrored in the vastness of the Sepik, the colors rippling across the spreading waters. I stayed overnight in a small village. Between the single line of huts and the river ran two parallel paths—one for women and only women, the other only for men. I tried walking on the women's path to see if anything would happen. A native man hurried over and angrily told me to move.

The next morning in another village a wide channel rolled up to a row of huts, carefully constructed atop supports. Marshes of pink lilies added color to the grey, weather-worn houses. This was the only village that I'd seen with cows. As the canoe approached, the animals watched and cleared a path, withdrawing to nearby safer spots. In contrast, Highland pigs will either freeze in fright or run along the road in front of approaching cars.

The cows mainly roamed around the houses and the *haus tambaran*, along with a few chickens and roosters, tolerating human presence. The cattle chewed the grass as if they belonged and men didn't. The *haus tambaran* sat comfortably at the end of a long, dense green fairway. The other houses sat twenty yards away. Meter-high mounds, small hills, carpeted with grass, lined either side of the lawn. From atop each mound swayed slender arms of coconut palms and banana plants.

The Sepik River.

Boats, houses, and people on the Sepik.

Children on the Sepik.

A crocodile on the Sepik.

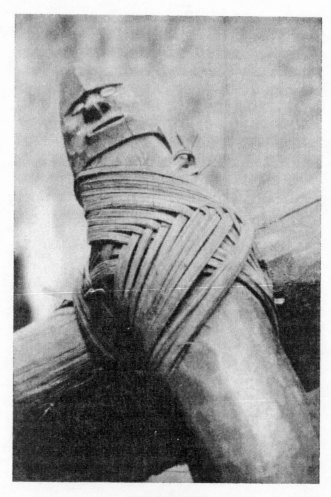

Sepik carvings.

Sepik masks.

Sepik carvings.

Haus Tambaran (Houses of the Spirits).

Haus Tambaran (House of the Spirits) with a sloping front.

Along the Sepik.

Next, we headed up a tributary to the Chambri lakes. Forests lined the side of the waters—now pure black. A breeze approached. A weed brought to New Guinea by a missionary in a fish tank that he had emptied out when he left had grown to the extent that it now clogged the waterway. The plant had thrived, and had created massive floating islands, in the centers of which other grasses and plants had taken root, joined later by ants and spiders. These huge floating carpets of life now blocked access to whole villages. Our canoe slowed to a halt, and we grabbed at the grass. "One, two, three . . ." We tried tugging ourselves toward it. I ripped up handfuls of the weed from its roots, but we didn't budge. The canoe owner pushed his oar along the river bottom and I dug my fingers deep into the islands. Together we yanked ourselves forward six inches. We pulled again and slipped by one island that had been blocking our prow. Gradually we squeezed through the rest. For months each year, the weed completely clogged all channels, then loosened to its present state. Eventually, the islands rotted in the sun, and sank to the bottom, turning the water black.

The villages, now isolated, sat at the side of the lake, containing beautifully constructed huts. At either end of every house perched white masks, placed there during ceremonies in which residents beat a special drum. The village makes clay pots decorated with a distinctive bird face. According to legend, the bird, a mythical creature named Gagu, consumed men. It ate a whole pig and had the pig's bones inside it. A woman tried to kill it with an axe, but couldn't. Two men eventually murdered it. As pottery, the bird still contains various things inside.

We turned back toward the river and headed toward the coast. I found a few older, less polished artifacts of anthropological interest, distinctive amidst those made for tourists. The villagers were surprised. "But that is *tumbuna!*" they said in astonishment at my picking it these prices alone out of a long line of artifacts.

"I know," I said. "I can tell." These works were older, carved with more feeling and power. I bought a hooked *bilum* holder with a woman whose head, viewed on the side, was that of a hornbill bird—perhaps expressing men's views of women. But I now regretted having passed up less tourist-oriented items upstream.

The sky became overcast. Sheets of grey obliterated all but a few glimpses of white and blue, as we headed toward the coast. The next village no longer had a *haus tambaran*—it had been destroyed by missionaries decades ago, and never rebuilt. The village still sold carvings, but they were without spirit or inspiration—the most boring and uninspired set of figures I had yet seen—showing the sad fate of a people lost without their culture.

The following village, built on a mosquito-ridden marsh, was an unofficial brothel. Canoes paddled up to mine to ascertain my interest. All negotiations, I saw, were conducted via canoe. As we neared the mouth of the river, men ever more accustomed to tourists nagged me, thrusting artifacts—as if they were babies and following me back all the way to my boat, which was docked at the far end of the village. The men swarmed around me like mosquitoes, and soon held out carved birds of paradise like offerings toward a feared god. The village's *haus tambaran* was crowded with newly made, uninspired masks in a wide range of sizes neatly hanging in rows, as in a supermarket or department store. The artifacts were now akin to mediocre imitation paintings of the great masters. I asked a man about true *tumbuna*. "They are in each man's house," he told me, quickly thinking of an excuse. This region had been contacted by outsiders earlier than the Highlands, and I was seeing the effects. It was a foreshadowing of how the Highlands might turn out in the future, I thought.

At the next village, children, men, and women suddenly appeared from behind trees and buildings, each carrying a folded sheet of dirty plastic under one arm and a clutter of wooden "souvenirs" in their other. They rushed to form a line, producing an instant bazaar. They laid out carvings and penis gourds which had not been used here for as long as anyone could remember. Artifacts were growing ever more similar and false, the villagers ever more thirsty to sell their copied goods to naive tourists. Mosquitoes besieged me in hordes around the scattered motionless puddles. In this malaria-ridden area, I soon counted 112 mosquito bites on my right arm alone. I now understood how undrained swamps bred malaria. My exhausted limbs sweat profusely. Sweat, roll-on mosquito repellant, oil from canned mackerel, juice from canned peaches, crumbs of a loaf of white bread, and mosquito bites all covered me. My feet felt like useless appendages, having laid limply in the canoe.

At the second to last village on the river, connected by a spur of the road back to Wewak, the *haus tambaran* had rotted away, and a new building had been built by South Pacific Beer (owned by Heinekin) as a bar, essentially replacing the men's house—once a place of worship. Cartons of empty beer bottles, in some cases having been reused to hold lime (chewed with betel nut), were stacked in the corners.

At the final village, on the main dirt road—a potential day trip from the coastal town of Wewak with its airstrip—newly carved artifacts for potential tourists stuffed a large new *haus tambaran* that sold beer as well.

The road from Angoram back to Wewak rumbled straight through the jungle. Midway, colored butterflies—some with six-inch wingspans—danced in open spaces around vines. They seemed oblivious to the growing encroachment of the outside world. Their days were numbered.

In Wewak, I stayed in a modern, Western-style hotel, beautifully situated on the Pacific. The sun lowered over the peninsula in luminous yellows as the waves washed over the shore.

When I checked in I was surprised to find a message waiting for me. "Call the Institute in Madang before you arrive," it read. I didn't even know they knew where I was. But this was the only hotel in town.

The hotel manager was a heavy-set Australian, and all the employees were native, and servile and submissive, lending a colonial air to the place. Only a handful of guests were staying at the hotel. I had a drink at the bar and met an Australian guest. "I hate Highlanders," he told me as soon as he heard where I'd been working. "But I like coastal people. I also now understand why South Africans refused equal partnership with nationals." Indeed, it was always the most racist people who, as if to deny their racism, insisted on calling natives "nationals" or "locals." "I understand that they have not had enough contact with Westerners to have accepted our ways," he told me. "But they still seem arrogant and too eager to fight. Even those who have been exposed to Whites—the natives working in my office—become lost when they relax and drink beer. They don't have the capacity for relaxed social conversation."

"But I think that is because they are used to dealing with *wontoks*, not strangers."

"No. It's just the way they are."

The next day I tried to catch a boat back to Madang. I made calls, but it was 1:15 by the time I found out when the boat left for Madang— at 2:00. I would not be able to catch it, since I was without a car.

I'd have to chance the airplane in the morning. A plane left once a week, and I didn't have a reservation. I would have to fly standby, and might not make it onboard.

I arrived at the airport at 7:00 a.m. to try to get on the 11:00 a.m. flight.

I was ready for a vigil. Anguish and pressure swelled up inside me. Finally, the petty officials behind the desk started calling out the waiting list. I wasn't called—although I had been the first passenger there. "Why haven't I been called?" I asked.

"These other passengers had reservations," he said. I suspected a lie. Why were they waiting to be called if they had reservations? I approached two. "Were you on the waiting list?"

"Yes."

"Did you have a reservation?" They looked at me as if I were crazy. "No."

I approached the clerk again, but he tried to avoid me. "Inside, inside," he said nervously, "they're making seat arrangements inside. I'm just checking them in."

"Fine. Then I'll get checked in as well," I said to the other attendant. He said to place my bags on the scale, which I did. There they sat. I then asked whether my ticket was finished. It was then prepared. I would get onboard, though the chaos around me continued. There was no system, and no one seemed to mind. The fact that there was an airplane and an airport at all seemed somehow miraculous. No one was used to or expected order.

I called the Institute at Madang and told them when I'd be arriving.

Up in the air at last, I again saw the scattered beautiful green islands of Madang ringed with scruffs of white beaches. I was anxious to see Sarah and my other new friends there. I was glad to have visited the Sepik, though disturbed by it, too, by what had happened there. These villages, more accessible because of water, were a few decades ahead of the Highlands in their contacts with the outside world. It was a foreboding glimpse of the future.

The Island that Ate
the Girl

When I landed back in Madang, Brian was waiting. He was quiet as he took one of my smaller bags and led me to his car in the parking lot.

Something seemed wrong. "What's up?" I asked.

"You haven't heard, have you?"

"No."

We sat down in the front seat. He looked down at the steering wheel, and didn't start the motor, for what seemed like several minutes. "Sarah," he finally said quietly, sighing, "is dead."

"She's dead?" I was shocked. His words sounded unreal.

He nodded slowly.

"What are you talking about?"

He kept just staring blankly at the steering wheel. "She drowned," he finally muttered. "On Karkar." We sat there for several more minutes, the runway beside us now silent and empty. Finally he continued: "We took the boat there last Saturday as we had planned. It took us six or seven hours and we arrived at dusk. In the morning, it was drizzling lightly. The villagers told us we shouldn't go up the mountain. Especially, they said, since when rain falls, the lava flows get extra slippery and fill up with water fast. But we thought the natives were just being superstitious, and we decided to proceed anyway. We left the base camp village to make the ascent, but after a few hours the rain thickened. Still, we kept hiking. Finally, we came to a point where two beds of lava crossed. By now, rivers of water were gushing down along each, and then joining together."

"A few of us started to cross. But then the rest of us thought that it looked too dangerous, and that we should head back. Three had already gone ahead, including Sarah. I went to fetch them, and was able to catch up and tell them about the decision to trek back. We turned around, going along very carefully, one step at a time, holding hands in case one of us slipped. We crossed one of the forked streams, and made it to the middle. Then we started to cross the other channel. But by now the water had swelled up even more, and was even harder to ford. I got across, then Frances, and then Nick. Sarah was the last one. As she reached up onto the final bank, her foot slipped, and she fell. Her hand let go of Nick's and the rushing water swept her away. The rest of us chased after her, but a little ways downstream we came to a thirty-foot waterfall. She wasn't at the top, and we hurried down along the side, but didn't see her at the bottom either. We kept walking, following the waterway, well into the night and early morning. By then, natives from the village had joined us. We got to the base camp, but still hadn't found her."

"In the morning, the natives discovered her body. She had practically reached the coast. She had hit her head—probably right away. The natives contacted the local mission, which owned a short-wave radio, and arranged for an airlift to get us off the island. But by then it was too late."

"Frances left yesterday to take her body back to Melbourne. She borrowed that book you were reading, by the way—Proust, I think— and took it with her. The natives are saying that the island got angry— because we were hiking up its back—and ate her. They have been making their own prayers for her spirit, and to appease those of the mountain. They have also forbidden anyone ever to hike up to the top of the mountain again. They say it is a place *bilong maselai*."

Brian and I drove back the rest of the way in silence. At the Institute I went up to my room, dropped my bag off, and walked outside. I looked at the lawn chairs where I had sat with her, beneath the palms and distant mountains. I felt disbelief, numbness. Her death touched me as those of kuru patients hadn't. I knew her as more of a person—as a friend, not a patient. I realized, too, that it could easily have been me. I had considered joining the trip to Karkar, and could have drowned. I myself had crossed dangerous rivers in the Highlands. I might not survive my stay either—none of us might.

I realized how much native beliefs had a role and rationale of their own, even if not always clear. I thought about what had happened to Michael Rockefeller, who had drowned when he had tried to swim ashore after his canoe capsized at the mouth of the Eilandon River in the Asmat area, on the southern coast of West New Guinea. The natives had warned him beforehand that if he were to capsize, he should hold onto the canoe and not try to swim to land, as the currents could be fierce, even if the surface looked calm. But he had ignored their advice. Some things natives knew better than we did—not everything, but some. We needed at least to listen.

Life in this country was precarious. Were the dangers worth it? I now believed they were—even more so as a result of what had happened to Sarah. I saw how short our lives were and how important it thus was to keep doing what we wanted to do and to live as fully as possible. But I was frightened, too.

Brian had started playing wheelchair basketball with Sarah's patient at the hospital. Brian seemed to take her death very hard, and was assuming a lot of the responsibility for making sure things got done since the accident. He seemed somehow to blame himself for what had happened.

Though few had known her, word about her death spread rapidly through the expatriate community in Madang and elsewhere in New Guinea. I later learned that Sarah's parents and friends back home also set up a fund in her name at the Institute to carry on her work there.

A few days later, my time came to leave and return to the Highlands. Brian drove me to the airport in the late afternoon, and we quietly said goodbye. I didn't know what else to say or do. We promised we'd stay in touch, and I hoped that we would. (In fact, we never did.)

The flight back was cloudy. From the plane, I looked behind me at Karkar, now covered in mist. Huge clouds charged forth from the sea—their heads bent forward, their bodies lagging behind. I found my lips working to utter the Sh'ma—the shortest but most important Jewish prayer. I was surprised it came to me—the only one I knew.

Coming of Age

When I got back to the Highlands, it was now "time bilong sun." Weather records show that the rainy season wouldn't end for another month. But none of the villagers had calendars or attached much significance to the passing of particular months except when coffee was harvested, just before the rainy season started. Thus, the belief that "time bilong sun" had started meant that it had. Everyone now stayed outdoors more, especially in the cool, late afternoons when the sun swam below the mountains in a final flourish of orange, red, pink, and purple streaks. Daily storms ceased; the mood in the village lightened. The power of the mountains uplifted me.

In Madang, memories of Waisa were small and faint, but they came back quickly now, even the pinewood smell and pleasant seclusion of the outhouse. The flowers I had helped Maryanne plant in our garden had all grown surprisingly large and fulsome in my absence. Sayuma seemed even more scheming for material advantage than I'd remembered, and much more nervous, carving his fingernail in the wood as he spoke to me, and starting the conversation by commenting on how much the passion fruit had grown on the porch. Admittedly, they had. "Sana hasn't been round," he was quick to tell me. Busakara soon came by, selling me a handmade *bilum* from Awarosa. He wanted 10 kina for it. I bid him down to a much more reasonable 5 kina, though I knew it would cost only 3 kina along the Sepik. "And I still have your raincoat marked," he reminded me as he walked away.

Inside, I unpacked. The house felt comfortable. At noon, Hazel knocked at the door. "Look what I got in the mail here." She showed me a fat packet of scientific reprints Carleton had sent her. "They're obviously not for me," she said. "But for you." Carleton had added her to his mailing list.

I got additional information on the possible cluster of patients I had recognized while at Madang. I was right. Mabamo and Paul, two male cousins, had both attended only two feasts—those of their father's sisters, Anero and Togawa, in 1953 and 1954. Anero participated in Togawa's feast and then died one year later. Paul's *nagaiya* or age mate, Timani, had died of kuru twenty years ago. Timani's sister, Kamari, was still alive. She came and spoke with me, along with several older men and women. "What happened when Anero died?" I asked. Kamari hesitated momentarily. "This is important," I said, "because it helps us understand how kuru was spread, and how it affects people."

One young man in the crowd, who had been on the coast, explained that I spoke the truth.

"Anero was cut and cooked," she said. "But after this feast, a patrolman was living in Bamusi hamlet, and pressured us to stop. Several men insisted that the next person who died of kuru, Kandabi, not be eaten. The corpse was buried. But at night a group of women secretly dug up the body and ate it." Clearly, the practice was already on the wane. Anero's was the last feast that the village fully accepted. Three boys were taken to it—Paul, his cousin Mabamo, and Timani. Paul and Timani were still being breast fed. Timani's mother, Amanali, was the only adult who slept in Anero's hut, and took care of her as she declined. These three boys' mothers were among the chief mourners and most active participants in the feasts. Paul and Mabamo displayed initial symptoms twenty-four years later, and both died a year afterward.

"Did they attend any other feasts?" I asked.

"No. There were no others in their line."

Paul and Timani, though born within a few days of each other, had widely different incubation periods. I went down the list of others in their kinship line. Fourteen had participated in Anero's feast, ten of whom later died of kuru. Two died of other causes. Two were still alive—both infants at the time—Kamari, and Imenaba, who also came by and met me. Both of them looked healthy. Kamari had just as much chance as Timani of being exposed, but was still alive. There was no doubt she'd been touched by her mother's infected hands. (As was the custom, I was again told, her mother hadn't washed her hands for a few weeks following the funeral.) But not everyone who comes into contact with an infectious protein gets infected.

I looked in the epidemiological record: fifty-one cases were listed from this village. I asked about twenty-one of them. Thirteen were at Anero's feast, seven weren't—three of whom were from Ilesa village and married here later, one was from Takai, another from Paiti, and two were born afterward. Only about one did nobody know for sure. Though the three boys had died after having attended only one feast, one had died at age 8, and the other two at age 28. The two boys, exactly the same age, died twenty years apart. In other words, age did not determine incubation period. Nor did the strain of the agent, as these boys had both been to only that one feast in their lifetimes.

I had also made arrangements with Michael in Goroka to have blood drawn from the elderly women I had identified. Ray Sparks drove out to Waisa one afternoon to draw the blood. He and Jeanie McKenzie, who accompanied him, both worked at the Institute. The next morning, the three of us set out. Jeanie got behind the wheel.

"Women don't drive cars!" Sayuma insisted, upset.

"Why not?"

"They can't."

"Sure they can."

"No they can't!" He was shaking. But Jeanie wouldn't budge. She was enjoying confronting his male chauvinism. I didn't like to see Sayuma hurt, but such tribal beliefs bothered me, too. Jeanie stayed behind the wheel, though Sayuma was still beside himself. We drove on to Takai. There, Ray took blood samples from two older women I had previously identified—Taniya, who was in her 70's and Anasuga, aged 55 or 60. He also sampled two controls—Tali, aged 35, and Tpeegapala, aged 65. We then drove further down the road, parked the truck, and walked over a rocky ridge to Agusatori hamlet, where he took a sample from Sapaya, aged 70.

Back at Waisa, we chopped wood for the evening's fire, and then dined on roast chicken and wine they had brought out from Goroka. Ray spoke in a soft, raspy voice of days past in Papua New Guinea. The fire crackled, radiating warmth as the black night folded about us, with only a few crickets disturbing the resounding silence. Ray sported a rumpled, soft woolen Irish tweed hat, its rim rolled up above his eyes. His whitening beard, blue sparkling eyes, and rounded hat perched on his reddish brown hair lent him the distinctive, free-spirited mark of the naturalist—a true Konrad Lorenz. A certain aura lingers around such individuals; they are at one with the natural world itself.

Roger stood up and began clearing the table. "What else do you do out here?" Ray asked. Roger and I looked at each other, unsure what to say. "Are there any card or board games?"

"There's a Scrabble set," Roger said as he walked into the kitchen. "Where is it?" Ray asked.

"It's just inside my room, on the right of the door," Roger called out. "Help yourself." Ray looked at me and I got up to get it. I realized it was the only time I had been in Roger and Maryanne's room. It had felt off-limits, and I had tried to respect their privacy as much as possible. Now, shortly before I was to leave for good, I was allowed in.

By the fire we all played Scrabble, and talked of films. Ray liked *Women in Love,* which drew him to D. H. Lawrence. Jeanie said her favorite was an Italian director whom we may not have heard of— Bertolucci. One assumes nothing out here in the bush.

The next morning we drove to Purosa, picked up Sana, and continued on to Mugiamuti, where we waited for two elderly women, Orika and Wanumemba—"old dollies" in Ray's words—to come down. We presented our case to them. The crowd listened, and the men conversed among themselves. An older man, with a big, greasy smile plastered on his wrinkled face, finally spoke up. Translation: he was one woman's husband and the other's son (which seemed an impossibility, since they were almost the same age). He would let us take blood if we paid him kina.

"We are taking blood to help the people's health, not for our own selfish gain," Ray said.

"You have forks and knives," the man replied, turning to us. "We must eat with our hands. Our hands are all we have. The White men brought kuru here. Before airplanes came, there was no kuru." Ray repeated his arguments, to no avail.

"Shouldn't we tell him he's wrong?" I whispered to Ray. "Many people died before White men came."

"No. Don't pressure the people. If we force them to give us blood this time, they will adamantly refuse next time." His was a long-term perspective: he would still be here—I would not. I deferred to Ray and we left. The man who had used a double standard and lied to frustrate us was still smiling.

Drawing blood. (Photo courtesy of J. McKenzie).

At Purosa, Sana was helpful and brought down his "mama," who appeared to be about 70, to give blood. Sana's wife gave me a *bilum* she had made, along with a bowl of tomatoes, a pineapple, bananas, and onions. Sana wanted to go to Goroka and was anxious to get his mail, presumably waiting to hear if Carleton would send him to America with me. I had explained to Sana that I was not going directly home, and that I thus couldn't take him, but he didn't seem to understand. He had been helpful and kind, and it was good to see him again, though he talked to me only about his "business" interests: my taking him to Goroka and his wanting Roger to come and install his tank ("The chickens need water").

I drove away with Ray, Jeanie, and Sayuma; we stopped on the mountain above Waisa for a picnic lunch.

Ray's blue eyes exploded in ecstasy as he spied a butterfly. He grabbed his butterfly net—he had brought three—and stalked the un-knowing velvety black creature, which he caught in a single swoosh. Its wings had brilliant yellow swaths, with four small dabs of matching gold balanced delicately at the edge. On the back, four fine, pale blue stripes were etched deep in the black fuzz. Ray calmly recited family, genus, and species to Sayuma, who nodded but clearly didn't under-stand a word. Ray was a biologist, not an anthropologist. As he spoke, another butterfly fluttered cross his field of vision, flitting through col-ored flowers a few yards away. Ray leaped up, and galloped after it, his white silk net bouncing and his hand grasping his hat, looking like the Mad Hatter. He dashed off along the green foliage by the side of the road and caught the insect. Later, he snatched a purple and green bee-tle, "sacred to the ancient Egyptians," and another long, flying beetle—light grey with flurries of black spots and a tint of orange etched along the edge—that had huge, round, textured black eyes, and fat mandibles that made a rhythmic croaking, opening and closing.

We drove back to Waisa, and Ray and Jeanie left for Goroka. Ray took with him *Current Contents*—synopses of the world's major sci-entific and medical journals—which he read, even as he sat in the car waiting.

Two weeks later, my last night in Waisa arrived. In the morning, I would depart for good. I packed, folding my few clothes. When I came to my poncho, I laid it aside, thinking of Busakara. I picked up my

boots to squeeze them into my lu;
by the side of my bed. I could decic
to Sayuma. Roger cracked peanuts
fussed over dinner in the kitchen.

I had mixed feelings about de
short. Much of it I had rushed. Yet
found out more about the infectic
nealogies on sixty-five patients, an
determine rigorously when exposui
natural incubation periods in hun
been done before (nor would it be pussible to as the
would know exactly when exposure for any case there had occurred).

I had found three clusters—with the patients in each developing
kuru virtually simultaneously after having been infected at the same
one or two feasts, occurring close together in time. The three pairs had
incubation periods of twenty-one, twenty-four, and twenty-eight years,
and didn't vary by more than a year between each member of the pair.
The disease can thus follow a uniform course of incubation (presumably a uniform "doubling time" of the agent) in two or more people,
even when the incubation period is over two decades. Yet some participants, such as Timani, had much shorter incubation periods. Age and
viral strains therefore did not determine incubation period. Perhaps, I
thought, the initial dose of the agent or the genetics of the infected individual did.

Yet as the number of current patients was small, these long incubation periods were the exception rather than the rule—at the far end
of the bell curve. Of over twenty-five hundred kuru cases since 1957,
only fifteen had incubation periods of five years or less. I had now
showed that the number at the high end was also small, paralleling the
behavior of these diseases in lab animals—only a few of whom develop
the disease much earlier or later than the rest. Thus in Britain now, we
will not have to wait thirty or forty years to see how many people will
die from Mad Cow disease. Most who will die will do so much sooner.

I had also found that almost fifty people were present at these
feasts. Thus people had attended many feasts, and had many opportunities to become infected in their lifetimes. Moreover, participants did
not wash their hands for a few weeks after the funeral, and could thus
expose themselves repeatedly over that time and by many routes of

infection (by mout
mosquito and fl
rubbed their
very effect
tivity a
survi
th

n, mucus membranes, conjunctiva). They scratched
a bites—of which I now knew there to be many. They
eyes. In animals as well, oral consumption isn't always
ve. Ingestion of the pathogen by mouth has variable infec-
d range of incubation periods. In fact some people, like Taniya,
ved. Anasuga was also still alive, as were two of her six children,
ugh the other four had died of kuru. She must have been intensively
exposed. Infants also survived—Kamari and Imenaba. Most people had
died after repeated exposures. Yet not all died who were in contact with
the agent only once or twice. In Britain, the next hamburger won't kill
you. But many infected hamburgers eaten over many years may.

Contrary to the speculative theory by a few anthropologists that
cannibalism never existed (reported on even in an April 1997 issue of
the intellectual journal, *Lingua Franca*), these tribespeople from different villages told me about their brothers, sisters, and children consuming their mothers, fathers, aunts, and uncles—matter of factly, without
embarrassment, accusation, or excuse. The explanations for why some
didn't participate were clear and consistent—mostly due to living elsewhere at the time or not having yet been born—never because of abhorrence of the feasts. The details were also consistent (e.g., the feasts
were held in the deceased's gardens).

Despite native insistence otherwise, the numbers of patients continued to fall, and no one born after cannibalism stopped had kuru.
None of the claimed kuru patients born after 1959 whom I saw in fact
had the disease. Kendabi's consumption was controversial within her
village, and hence the last such sanctioned event there. Thus, cannibalistic feasts were indeed the cause of transmission, despite persistent
theories that they never existed

"Shouldn't I fight these theorists?" I later asked Carleton. "Argue
back in an anthropological journal and tell them how wrong they are?"

"No," Carleton said. "Don't get embroiled in the debate. Armchair
philosophers will doubt everything. If you show them photographs and
films of cannibalistic feasts, they'll claim the scenes were just posed or
acted. Merely document your scientific findings." I did. I remembered
that some academicians have even claimed that the Holocaust never
occurred in Europe and has just been made up.

Yet I had learned other things here in New Guinea as well.

boots to squeeze them into my luggage, then put them on the ground by the side of my bed. I could decide later whether or not to give them to Sayuma. Roger cracked peanuts over a book at the table. Maryanne fussed over dinner in the kitchen.

I had mixed feelings about departing. My time here now seemed short. Much of it I had rushed. Yet I had accomplished my goals. I had found out more about the infectious agent. In all, I had collected genealogies on sixty-five patients, and had showed that it was possible to determine rigorously when exposure occurred, thus calculating precisely natural incubation periods in humans after a meal—which had never been done before (nor would it be possible to do in Britain, since no one would know exactly when exposure for any case there had occurred).

I had found three clusters—with the patients in each developing kuru virtually simultaneously after having been infected at the same one or two feasts, occurring close together in time. The three pairs had incubation periods of twenty-one, twenty-four, and twenty-eight years, and didn't vary by more than a year between each member of the pair. The disease can thus follow a uniform course of incubation (presumably a uniform "doubling time" of the agent) in two or more people, even when the incubation period is over two decades. Yet some participants, such as Timani, had much shorter incubation periods. Age and viral strains therefore did not determine incubation period. Perhaps, I thought, the initial dose of the agent or the genetics of the infected individual did.

Yet as the number of current patients was small, these long incubation periods were the exception rather than the rule—at the far end of the bell curve. Of over twenty-five hundred kuru cases since 1957, only fifteen had incubation periods of five years or less. I had now showed that the number at the high end was also small, paralleling the behavior of these diseases in lab animals—only a few of whom develop the disease much earlier or later than the rest. Thus in Britain now, we will not have to wait thirty or forty years to see how many people will die from Mad Cow disease. Most who will die will do so much sooner.

I had also found that almost fifty people were present at these feasts. Thus people had attended many feasts, and had many opportunities to become infected in their lifetimes. Moreover, participants did not wash their hands for a few weeks after the funeral, and could thus expose themselves repeatedly over that time and by many routes of infection (by mouth, mucus membranes, conjunctiva). They scratched mosquito and flea bites—of which I now knew there to be many. They rubbed their eyes. In animals as well, oral consumption isn't always very effective. Ingestion of the pathogen by mouth has variable infectivity and range of incubation periods. In fact some people, like Taniya, survived. Anasuga was also still alive, as were two of her six children, though the other four had died of kuru. She must have been intensively exposed. Infants also survived—Kamari and Imenaba. Most people had died after repeated exposures. Yet not all died who were in contact with the agent only once or twice. In Britain, the next hamburger won't kill you. But many infected hamburgers eaten over many years may.

Contrary to the speculative theory by a few anthropologists that cannibalism never existed (reported on even in an April 1997 issue of the intellectual journal, *Lingua Franca*), these tribespeople from different villages told me about their brothers, sisters, and children consuming their mothers, fathers, aunts, and uncles—matter of factly, without embarrassment, accusation, or excuse. The explanations for why some didn't participate were clear and consistent—mostly due to living elsewhere at the time or not having yet been born—never because of abhorrence of the feasts. The details were also consistent (e.g., the feasts were held in the deceased's gardens).

Despite native insistence otherwise, the numbers of patients continued to fall, and no one born after cannibalism stopped had kuru. None of the claimed kuru patients born after 1959 whom I saw in fact had the disease. Kendabi's consumption was controversial within her village, and hence the last such sanctioned event there. Thus, cannibalistic feasts were indeed the cause of transmission, despite persistent theories that they never existed

"Shouldn't I fight these theorists?" I later asked Carleton. "Argue back in an anthropological journal and tell them how wrong they are?"

"No," Carleton said. "Don't get embroiled in the debate. Armchair philosophers will doubt everything. If you show them photographs and films of cannibalistic feasts, they'll claim the scenes were just posed or acted. Merely document your scientific findings." I did. I remembered that some academicians have even claimed that the Holocaust never occurred in Europe and has just been made up.

Yet I had learned other things here in New Guinea as well.

I had seen the importance of culture in views of and approaches toward disease, treatments, epidemics, doctors, and death. I had learned most about these particular aspects of the Fore world view, but they reflected the coherence and the internal logic of the tribe's beliefs about a world filled with spirits and magic. I had seen how pervasively the epidemic effected the Fore's everyday life—their fears about my laundry and their possessions. Yet much of life here now continued on as if nothing had ever happened—people still ate, slept, worked in their gardens, played cards, talked by their huts at night. Once an epidemic has begun to wane, life in many respects returns to normal, though myths, fears, and memories persist. I seen a range of differences between societies—Highland and coastal, New Guinean and Western—had observed closely Pacific islands, rain forests, mountain ranges, and coral reefs, and come to understand more about primitive art and life. I had written my first scientific paper, and learned firsthand how to do science—how it worked and how much of research occurs in small steps and with difficulty.

I had come to feel more confident and sure of myself, having undertaken a difficult project, and carried it out. I had come to trust more and stand up for my own instincts, hunches, and desires.

But the work here had not been easy. I was no longer the romantic idealist I had started out. I had seen PNG as a paradise—men and women unspoiled by the West—but it wasn't. It never had been. This was not Adam and Eve. Materialism had existed here from the start. Despite Marxist theories about returning human beings to a less selfish state, greed was a basic element in Stone Age life. For millennia, idealism in the East and West had sought to avoid or change that trait. Christianity saw avarice as part of original sin. Greed created problems for religions that tried to transcend cultures—Christianity and Buddhism extending beyond boundaries of nation and race. But here greed was innate, and had to be acknowledged and accepted as part of human nature, revising moral and political expectations and goals. It lurked deep in the psyche, no matter what the society—still worth fighting, but ever present. The West was hardly immune—witness world wars and Wall Street. "Man was born free, but everywhere is in chains," Rousseau had written. Yet mankind was also born scheming, and remained so everywhere he lived. Man was difficult not just because of society's "corruption."

Here on the equator, conditions were harsh. Darwin's most important contribution was undoubtedly the theory of natural selection; and in the survival of the fittest, psychological warfare plays an important role. The jungle bred warfare. Human beings fought parasites and other human beings. One culture battled another; science, superstition. Many Westerners have left PNG, bitter after years. In the end, despite Westerners' dedication, natives wanted them out. The land did not belong to foreigners, and outsiders left feeling unappreciated and used. Enormous disparities between us remained.

Highlanders wanted Westernization, though I had seen its potential dangers along the Sepik. I feared that Western development would hurl the primitive Fore forward more destructively and uncontrollably than they could ever guess. The tribesmen weren't "pure," though that didn't make the West less "contaminating." Modernization offers many benefits. It should be encouraged, not prevented. What a shame, however, that the bad rushes in with the good. I hoped development could proceed with care and attention paid to its potential side effects that, if addressed, might be ameliorated. But it wasn't clear that this would occur, or how, especially given political corruption. I was left troubled.

I had also seen how humans are part of a larger biological continuum. Rocks had supported lichen—among the earliest life forms. Evolution had spawned other species and eventually homo sapiens, who had moved from the Stone Age to the Space Age. Yet competition for resources had continued. Capitalism, which I had disliked when embarking on my trip, was but the most recent manifestation of a long process that began with the moss atop Mount Wilhelm.

The other most recent developments in this on-going biological progression were epidemics. Parasites battle against their hosts and prowl for new niches. As William McNeil wrote in *Peoples and Plagues,* and Laurie Garrett in *The Coming Plague,* epidemics result from technological advances—from sailing vessels to flying machines. Kuru and Mad Cow disease are but two examples of how biological and social mechanisms interconnect.

Biology shapes us more than we generally think. I had observed how personality types in a village don't vary much between societies. I had seen Sayuma's slyness and Sana as a country gentleman farmer and family man. Yet to view personality as biologically based contrasts with Freud, for instance, who posits intrapsychic conflict and experience as the root of who we are. Nonetheless, character may have caused the conflicts, not vice versa. Lies and psychological gaming cut across societies. To fail to see man and nature as linked is merely a residue of the medieval Great Chain of Being that cast us halfway between God and animals. In fact, genetically, we share more with animals than we differ from them. Our DNA code differs by less than 2% from that of the chimpanzee.

Each of us is an experiment of nature—a new product of our species, an attempt of that species to survive socially or biologically. Individual variation allows for adaptability, trial and error. We cannot predict the results—whether we will achieve what we set out to, what we will find, what difficulties we will encounter, where our interests or passions will lead. These outcomes depend on the environment into which we find ourselves; and chance. Yet nature works through us all.

Still, these commonalities get ignored. Genuine understanding between societies is much harder than I thought. Cultures pride themselves on their differences, each seeking superiority over others. It is difficult to surrender agendas and needs. I thought other New Guineans and Westerners would share my views and goals and support the scientific work I was doing. But they often did not.

As a result of these experiences, I saw myself more as American, as Jewish, as a fledgling medical and social scientist, and as a cultural observer. Here differences became more apparent. I had seen, too, how much landscape shapes cultures and individuals; and how much the United States and institutions in which I had worked had affected my own world view, but had limitations.

Here in New Guinea, I had come of age.

PART IV   RETURNS

Culture Shocks

I was ready to come home. But my return to the United States proved far more trying than I had thought. I experienced culture shock as I had when going to New Guinea in the first place.

At times in New Guinea I had felt like Dorothy in *The Wizard of Oz*, who had learned to appreciate home more after journeying to a far-off country—a mirror image of her own society in which characters were similar, played by the same actors. There, having encountered good and bad witches, she came to view home differently.

In New Guinea, I had decided definitely to attend medical school. I wanted to observe, work with, and understand people firsthand. This seemed what I was destined to do. Yet the bush had affected me far more than I had thought.

Over the subsequent years of medical school, internship, and residency, I saw how training often focused narrowly on medications, and paid scant attention to patients' lives outside the hospital or clinic—at home. I saw how little could be done medically for many patients. I spent hours with those who had rapidly advancing diseases, with the sole hope of, at best, slowing somewhat the illness's progression. At one point I cared for a patient named Charlotte Rowe who had Creutzfeldt–Jacob disease. She lay in a modern hospital, where her daughter visited daily. But Charlotte didn't recognize her daughter or anyone else. As a physician, I could do nothing more for her than Sila had done for his patients with kuru.

In psychiatry particularly, limited knowledge of the mind and the brain led to views of and approaches toward illness that reminded me of those in the Highlands. As a resident in psychiatry, I treated a patient, Cynthia Thomson, with Freudian-based psychotherapy. She became depressed while struggling through medical school. I told my supervisor, a strict Freudian, that I wanted to offer more supportive treatment. "No," he said to me. "Just say nothing to her."

"Nothing?"

"That's right, get her to free-associate."

But when I said nothing, Cynthia became increasingly frustrated, and started coming late and missing sessions.

"She's failed the treatment," my supervisor suddenly announced one day. "Terminate her."

"Terminate her?" I asked. I felt uneasy. The treatment seemed to have failed her—not vice versa.

"Maybe the approach wasn't right for her," I suggested.

"No. There's nothing wrong with our approach. She's just a no-goodnick. She's just resistant, not psychologically minded." I was amazed. Sila didn't either consider that his treatment might be ineffective. The patients were different in each case, but the responses elicited were similar: the patients were blamed for the treatment failures. Through medical school, internship, and residency, I looked at medicine and psychiatry with a different eye. I had assumed that medical science sought to uncover the truth at all times. But claims of truth now left me wary. The West has advanced far in many regards—but in other ways, not. I saw how we might be mistaken in our basic assumptions and never realize it or want to. Humans look for certainty in their beliefs, even if the beliefs are wrong.

I had thought the natives "primitive." Yet in the worlds of medicine and psychiatry, I saw many similarities to how New Guineans had understood and dealt with disease and the world. New Guinea gave me bases of comparison for all else I thought and saw—not only in medicine and psychiatry, but more broadly as well. My experiences there freed me somewhat from the constraints of Western culture: I took less for granted, and accepted less as given (which proved both a benefit and a handicap). I saw how easily violence erupts, as when Busakara and others marched off with bows and arrows. I observed how inventions get taken for granted, as with the wheels made by Fore children, though adults in the culture had never discovered the wheel. Among the Fore, no word existed for what lay beyond the biggest mountain.

Humans confidently feel they know the world, though they miss much. Purely historical explanations of human events now persuaded me less than those that acknowledged as well basic underlying patterns of human nature.

The year I returned to the United States, the initial cases of AIDS were reported. As the AIDS pandemic spread, I saw how Western reactions resembled Stone Age responses to kuru. Both diseases are fatal. In the case of AIDS, too, fear spread. Those infected or suspected of being infected—including gay men and injection drug users—were shunned. A desperate search for cures ensued with new advances heralded, often prematurely, and new remedies sought, frequently at considerable expense. Widespread death became a cultural as much as a medical or psychological phenomenon. Pressures to change culturally sanctioned behaviors that had led to the spread of the disease—in the case of AIDS, sexual and drug use behavior, in the case of kuru, cannibalism—met with enormous resistance. I had been surprised that the Fore had continued cannibalism in the face of the epidemic and that they hadn't connected the disease's spread with the feasts. The notion of infectious disease is clearly not automatic, and in fact counterintuitive in both New Guinea and the West. I was surprised by many gay men continuing to practice unsafe sex despite the mounting AIDS epidemic around them. Behaviors involving sex, death, diet, and mourning are deeply rooted biologically and imbued with extraordinarily powerful meanings, and are hence difficult to change even in the face of a plague. Rather, individuals look for rationalizations to continue their established behaviors: the disease can't happen here or to us, and results from sorcery, can be reversed by counter-sorcery, affects only gay men who are promiscuous or don't know their partners; the government has experts who have solved the problem.

Disease prevention programs have underaddressed these issues. The U.S. government has spent millions of dollars to change HIV-transmitting behaviors, thinking education alone would be enough, and for many years paid no attention to the cultural contexts and meanings of the relevant behaviors. Consequently, prevention efforts failed to be as effective as they might have been. More attention to these issues is still needed, as enormous cultural and psychological resistance arises to changing entrenched habits.

My experiences in PNG taught me the need in medicine, science, and psychiatry to examine closely social and cultural contexts, implicit assumptions, and unproven theories—how illnesses and treatments get defined and framed. I saw how diseases aren't givens, but are constructed by cultures in different ways. In the West, infectious agents have been isolated, enabling physicians to differentiate between, for example, *Klebsiella* and *Pneumococcus* pneumonia, based on which of these two bacteria is responsible. But such distinctions don't exist in New Guinea.

Similarly, I saw that medicine consisted not merely of decision-making trees—as I was taught in medical school—providing definitive solutions to problems, but of events unfolding, shaped by settings, biases, and personal interactions. Yet physicians ignore these contexts and interpretations.

New Guineans dismissed my views and I could dismiss theirs, but the comparisons were revealing. I saw the need to keep an open mind. Yet the more I entered my new professions, the more I saw the absence of such an approach and the difficulty in medical institutions of maintaining a broad viewpoint.

New Guinea had taught me, too, about doing research—identifying phenomena to study, framing questions, gathering and analyzing data. I had learned the obstacles as well as the rewards of science and saw science as an on-going experience. I was less intimidated by the fixed authority of textbooks, data, and journals.

I had gone to New Guinea wondering whether to become a physician or study culture. In the end, I learned the value of bringing these two interests together. Indeed, in the years since then, when others have told me of being torn between two ostensibly conflicting career paths, I have seen the wisdom of integrating strengths—particularly as the world becomes more complex and specialized, and the need for synthesis ever greater. New Guinea taught me the value of looking beyond the limitations of single categories and standpoints.

My experiences there turned out to be what I have drawn on most as a psychiatrist—more than Freudian or postmodern theory—in trying to understand culture and mind. I came to understand psychiatry less as a Freudian or a psychobiologist than as an anthropologist. My experience in PNG made me part of a club of ethnographers. After seeing science and medicine in New Guinea as human activities, with very human sides, I thought differently about what I experienced in the United States—how other trainees and I were taught day to day to handle death, dying, mental illness, and a variety of ethical issues, and how individuals affected by the new HIV epidemic responded to death around them and in their own lives, and organized their worlds. I began to write about medical internship and psychiatric training, to explore how our own perspectives on illness and treatment are shaped, and reflect our own biases and world view. I also studied HIV-infected patients' views of their illness and their lives and the culture of the HIV community, or as some call it, "HIV-land"—how the epidemic can mold patients' identities, and moral and social worlds. These activities all followed directly from my research on kuru. I probably would not have done them otherwise. But I saw the need for bringing together medicine and psychiatry with studies of society, language, and culture—in short for a Cultural Studies of medicine, akin to such approaches in other academic fields.

The number of cases of kuru has continued to fall. The disease has now all but vanished from New Guinea. Unfortunately, the tensions between the Stone Age and the modern world have only mounted. Several times, Jason has written and called me, asking for money. I still worry about the future he and his country face, caught between two worlds.

Mad Cows

The paper I wrote in Madang got published, and cited in leading textbooks. When Mad Cow disease broke out, this research suddenly became important. I saw that the significance of particular research cannot be fully predicted. When I set out to do this project, it was just an interesting problem. I had no idea it would have implications for other epidemics only a few years later. Yet, I learned, this was the nature of science. Our choices and findings may have more far-reaching consequences than we suspect at the time.

Having conducted the study, I remained in close contact with Carleton and other researchers in this area from the NIH, working with them for periods of time, through the emergence of Mad Cow disease. In 1996, human cases began appearing. Yet it remained unclear how they had become infected—probably through sick cows, but conceivably as well through ground-up sheep and cow bones, used in fertilizer on vegetables, or fed to pigs and chickens who might become infected or excrete it themselves. This manure might then be used as fertilizer too. Because the incubation period can be years, many domesticated animals, though possibly infected, get slaughtered for human consumption before becoming symptomatic. Material from ground-up cattle carcasses also was also used in the manufacture of other household products, including cosmetics and pharmaceuticals.

Journalists and the public demanded to know immediately how many people had been exposed to Mad Cow disease and would die as a result, and when, and how long the world would have to wait to know if British beef was now okay. The answers depended on how long the agent can take to display symptoms and what determines the incubation period (strain, dose, route of infection, or host genetics)—questions I had gone to New Guinea to address. These issues are important, as there is still no clinical test to see whether someone has been exposed to, or is incubating the agent.

Indeed, reports in the press that the incubation period of Mad Cow disease can be twenty years or longer were based in large part on the work I did. This research thus became relevant in trying to decide whether those who got Mad Cow disease were infected before or after the British government adopted preventive policies—and hence whether current consumers of British beef were still at risk. Work on kuru was suddenly important, since new evidence on bovine spongiform encephalopathy (BSE) can't be obtained directly (people would have to be fed material from infected cattle and then studied).

I observed and participated in discussions about the possible risks. Yet I saw the limitations of scientists and government officials in making public policy about science. Most scientists assumed British government regulations would make the beef supply safe, ignoring the possibility of human error in carrying out policy or needing to enforce it. British leaders assumed that Mad Cow disease wouldn't spread to humans because, they argued, scrapie never had. Yet through the late 1970s, scientists commonly thought that the higher incidence of Creutzfeld–Jacob disease (CJD) in certain areas, particularly North Africa, resulted from the fact that people there ate a lot of sheep's brains. The notion that scrapie could pass to humans had been generally assumed. It should have been no surprise to the British—certainly not to British scientists.

Many criticized the British government's handling of the situation. Yet their shortcomings were no different than those of other groups confronting an epidemic. Denial, belief that it can't happen here, and resistance to changing old behaviors because of other, competing desires (e.g., protecting the British cattle industry) impede prevention. Scientists also unequivocally assumed that policy would be carried out fully and correctly. "There's nothing to fear," most researchers said. "The human cases undoubtedly represent episodes of exposure from before precautions were taken." Yet prevention policies at slaughterhouses were indeed poorly monitored and carried out. I saw how scientists and policy makers, in part because of public fear, failed to communicate known aspects of the agent (such as the fact that it can persist in the soil for decades and infect, for example, future flocks of sheep). Science and policy-making require two different kinds of thinking. Policy-makers don't deal well with uncertainty. They want a yes or no answer—whether this or that will happen, and whether to do this or that in response. The problem stems partly from inability to accept the risk of disease outbreaks. As human beings, we like to feel we are safe. It is easier to proceed in other matters day to day if we feel we do not face danger.

The parallels—biological, social, and psychological—between kuru and Mad Cow disease became increasingly apparent, given the uncertainty and nonintuitive nature of these newly discovered infectious crystal proteins and the resistance that emerges to taking necessary precautions against them. The British government wanted to think that its nation could continue to eat its beef safely, just as the Stone Age New Guineans wanted to believe they could continue cannibalism. Both New Guinea natives and the British government had difficulty imagining or accepting a disease that has long incubation periods. The consequences for both groups have been disastrous. In Britain, seven hundred thousand infected cows entered the human food supply. The British government then exterminated hundreds of thousands more. In May 1997, Prime Minister John Major's Conservative government, much criticized for its handling of the situation, was ousted. For the first time in 16 years, a Conservative prime minister has not been in power. Other issues contributed to his fall, but Mad Cow disease was one.

In 1997, after the initial human cases of cattle-acquired CJD appeared, many scientists became far more cautious in their estimations. I asked Carleton how likely a widespread epidemic was. "How should I know?" he now responded. "Risk is hard to evaluate or know how to handle. Meteors have fallen and killed automobile drivers. But does that mean all drivers should wear helmets? We don't know how to respond to low-level risks."

I had now observed closely three major epidemics: kuru, Mad Cow disease, and AIDS. In all three, people at risk feared the disease, but had difficulty understanding the disease mechanisms, accepting long incubation periods, and changing deeply rooted behaviors—eating humans or British beef, having unsafe sex or using dirty needles. Those affected—Fore, cows, gay men, injecting drug users—were in turn rejected and shunned. (Though prejudice against groups affected by HIV existed beforehand, it now rose to new heights.) Surrounding groups responded as a whole—geographically. Despite differences in scientific knowledge base, theories of disease, and cosmological beliefs, basic reactions were similar. In the absence of clear, visible evidence of infection, disbelief and pseudo-science persisted. In Britain, infected beef looked, smelled, and tasted the same, unlike when other parasites—such as mold— invade.

Drastically but too belatedly, the lessons of these diseases have begun to sink in. Recently, a friend came over from Britain—a petite woman—and ordered huge steaks everywhere she went, no longer eating any steak at home. Some groups at risk of HIV have begun changing their behaviors, yet thousands of new cases of infection occur every year in the United States and abroad.

PART V   POSTSCRIPT

Where I'm From

Periodically, I'd wonder about going back to New Guinea. Senior researchers would often mention to me how interesting it would probably be to return. But I was wary. I had survived it once—the malaria, the intestinal parasites, the physical dangers. I didn't relish facing these risks again. New Guinea was also far and took a long time and a lot of money to get to. Yet increasingly, I saw how important my experiences there had proved to be.

Then, in 1997 I received an invitation to attend an AIDS conference in Australia. I had never been to Australia; I wanted to go to the conference, and knew it was as close as I'd get in a while to PNG. I told Carleton I was thinking of going back.

"Be careful," he said. "It's gotten very dangerous." When he had returned recently he had helicoptered in and out rather than drive on the roads and risk being held up. Rascals posed more of a threat. In addition, the *New York Times* published a few articles reporting violence. The Prime Minister, Sir Julius Chan, had called in foreign mercenary troops to quell rioting at the lucrative mines on Bougainville—an island off the coast, owned by New Guinea. The mercenaries had been promised a share of the mines' profits in return. The Minister of Defense, appalled, had resigned, and a crisis ensued. Bitterness and rage tore through the country. Rioting broke out in the capital and elsewhere, and the government had called a curfew throughout the land. National elections were now being held.

Unsure whether to go or not, I called Michael Alpers, who assured me that the Highlands near Goroka were safe. "The most dangerous thing here will probably be getting stuck in the mud, driving out to Waisa, and having to shovel yourself out. The past two months is also the first time we haven't had a confirmed case of kuru," he added. "It's the end of an epidemic."

The End of an Epidemic. The *New York Times* and other publications had recently suggested that the end of the AIDS epidemic was at hand, too. I wondered what happened at the end of epidemics—how people responded. Here was a chance to see.

I decided to go. I consulted an infectious disease specialist in New York, who ordered shots against tetanus, typhoid, hepatitis, and cholera, and prescribed an antimalarial, Larium. "It has one very unusual side effect, however," she warned me. "It causes terrible dreams." I had never heard of this as a side effect of any other medication, but I had no choice.

I also contacted the New Guinea embassy in Washington to get a visa. A woman there said she would send me a package of forms to complete. But the forms took two weeks to arrive. As soon as I received them, I completed and Fed-Ex'ed them back with a note that I was scheduled to leave in two weeks.

A week later, I called the PNG embassy to find out the status of my visa. "What's your name?" a woman asked me. I told her and spelled it. "Just a minute." She put me on hold. "I'm sorry," she then said in a definitive voice. "We don't have your material."

"What do you mean you don't have it?"

"We haven't received it."

My heart began to race. My passport was thus now missing. "How are you spelling my name?" I asked, grasping for explanations.

She grunted. "FLITZMAN."

I corrected her. She put me on hold again.

"No, we don't have it here," she repeated. I called Fed Ex, who said they delivered it six days earlier. I telephoned the embassy again.

"What's your phone number?" she finally asked with reluctance. "I'll call you back."

An hour later, she rang. "Yes, we have it. We'll do it as soon as we can." Half an hour later she called back again. "We also need a copy of your airline ticket in and out of the country."

"It wasn't clear in the instructions that that was needed."

"We need it."

"Can I fax it to you?"

"Okay."

I faxed it. Half an hour later, she called again. "We also need a check for $10.25." It hadn't been clear from their form that that was necessary either. They wouldn't take a credit card number, and I had to Fed Ex them a check. Luckily I had one with me at work.

The day before I was to leave, I received the visa. This disorganization reminded me of my time there. Bureaucracy was a foreign concept.

I flew to Australia—twenty-two hours of traveling from New York—and attended the conference. Then I flew to Brisbane, and changed planes for New Guinea. The world now seemed a smaller place. In forty-eight hours I could potentially go anywhere in the world, even to the remotest New Guinea Highlands.

As I boarded the flight to Port Moresby, I saw my first New Guinea resident again—a young boy, his face like a little man's, accompanied by his father. The boy strutted freely, his arms swinging. He was small but had big eyebrows, and smiled with an innocence more than that of most American kids—reflecting, it seemed, a lack of feigning and pressures as in the West. Then I heard his voice—soft, singsong. He brought New Guinea back to my mind.

Yet the flight contained only a handful of natives, and two hundred Caucasians. A chubby, pale-skinned man with a red beard sat next to me. He put down his book—a well-thumbed adventure novel—and leaned over to me. "My name's Ben," he said, reaching out his hand for me to shake. "First time in PNG?"

"No, I was here for a period many years ago."

"So you're a former expat! Once an expat, always an expat." He was working on building a new airport in Port Moresby. "There's tremendous wealth in this country—copper and gold, coffee, copra, now natural gas. They're going to build a pipeline under the water to Australia, and ship gas and petroleum to Japan. But the money doesn't go to those who need it. One minister took his whole village to Cairns, Australia for two weeks for his son's 21st birthday. For the election, brand-new four-wheel drives have begun to appear in villages—to buy votes. Foreign companies strip mine the land, bribing local leaders who sell the land rights. The Japanese denuded Borneo to make chopsticks. The same thing will happen here. The *wontok* system is destroying the country from the top down. There's also no concept of maintenance.

The new airport has fiberoptic connections between the buildings. Every time I look at it, I think it's just going to fall apart in a few years. The airport has no navigational system—it keeps breaking down. The tarmacs are all in disrepair. You've heard about the curfew?"

"It's still on?"

He nodded. "Port Moresby is a city under siege."

A few hours later, we landed. "Welcome to New Guinea," the steward announced cheerily over the loudspeaker in an Australian accent, as if it were a destination like any other. I looked out the window. There was New Guinea—the airport, and then nothing—a few small, scattered buildings, but mostly hills, green-brown dirt, and broad shadows under clouds and the strong equatorial sun.

Ben leaned over to me again. "Now the problems just begin. We have to get through customs. Watch: for this large flight, they'll have only one customs man on." We walked down the plane's aisle onto rickety stairs that set us down on the runway. The smells of New Guinea returned immediately—the earth, the dust, the smoke. The plane was much bigger, whiter, newer, and cleaner than any buildings around. We walked through a rusting covered walkway into the terminal.

Employees stood around with nothing to do—*wontoks* of someone, I thought. In customs, there were different lanes—tourist, transit, and resident. Each passenger took five to ten minutes to get through. I was the only tourist.

"Where's the flight to Goroka?" I asked one of the employees standing around.

She pointed up and up and over with her finger—as if it were a humming bird—directing three quick movements in the air, showing me to go out, around the back, and in.

I was confused, but left the room, which was for international arrivals, and walked outside, and back into another entrance, which was for domestic departures.

Outside, a crowd swarmed around the door. Between the two doorways, women sat on the pavement, colorfully dressed, barefoot, the bottoms of their feet rough. They were weaving *bilums* from colored yarns. Babies lolled in their laps.

The area in front of domestic departures was walled off by a metal fence. At the gate, two soldiers stood in chocolate brown uniforms with machine guns strapped over their shoulders, holding hands. When

I approached, I took out my ticket and showed it to them. They unlocked their hands and let me through, then held hands again, to block others in the crowd from getting in.

"Where's domestic departures?" I asked, to make sure in the confusion around me that I was going in the right direction. The guard didn't seem to understand. "Goroka?" I asked.

He let go of his companion, put his arm around my shoulder and with his free hand, now holding a walkie talkie, directed me into the building, to the side. Then he returned to his post.

Inside, most people were barefoot. Most of the florescent bulbs were out.

Departures were once listed on pieces of paper attached to posted clipboards, but the boards were now all broken. I walked up to a desk and was eventually checked in. I then had half an hour to wait for the flight.

There were rows of plastic chairs, some broken. On the wall spread a billboard of an astronaut with the Earth tiny in the background. It read: "Space available." I doubted most people around me would understand. I sat down next to the one other Caucasian in the room. He turned to me. "So," he asked, "are you here for gold?"

"Gold?"

"Yes. The man just sitting here was. He's a German mining engineer from a firm exploring an island north of Madang." I explained I was here to do research.

I got up and went to the bathroom. The stalls lacked doors. The toilet paper hung six feet up, from the metal bar on top of the partition, connected to the ceiling.

Back in the waiting area, I felt hot and drained. Finally, my flight was announced. To board the plane we walked outside again. Another plane suddenly approached and landed. All of us—employees and passengers—stopped and covered our ears to block the deafening sound. The plane stopped fifteen feet in front of us.

Up in the air, the clouds differed from those I knew back home. A dense, white woolen blanket covered all, with occasional tall stalks, resembling mushroom clouds, bursting upward.

When we began to descend below the clouds, the Highlands suddenly appeared—rugged with no flat spots, just hills, seemingly unpopulated, entirely coated with thick trees. The mountains seemed higher and steeper than I remembered.

As the plane landed in Goroka, people ran over from the surrounding lawns and streets, gathering around the barbed wire fence that surrounded the runway, to watch. The ground crew was joined by a private security force wearing navy blue uniforms with yellow arm patches and yellow hard hats. We walked down a stairway and over to the small building—the Goroka airport. How odd, I thought, that I was last here 16 years ago. It seemed so far from my experience, from my life since, yet in other ways so close.

A ground crew member dragged the baggage to us from the plane on a cart by hand, and placed it on an aluminum sheet hammered onto a wooden table—all outdoors. He threw on the ground bundles of newspapers that had been checked in. As they banged, clouds of dust rose up.

When we all collected our luggage, a ground crewman unfastened the padlock on a chained gate behind us. Outside, a crowd surrounded us, mostly curious onlookers. The street was dirty and dusty, the air hot. I saw the familiar faces of Eastern Highlanders.

Deborah Lehman from the Institute met me. "I've made a reservation for you at 'The Bird'"—Goroka's one hotel, the Bird of Paradise. "A lot of people are in town, waiting for the election results," she said as she drove me there. Election posters—hundreds of them—plastered every tree. Thousands of candidates were running for little more than a hundred seats. In some areas, each village or kinship line had put up a candidate.

Four guards with hard hats and police clubs and a police dog blocked the hotel entrance—election results were being counted inside. She dropped me off, and suggested I come to the Institute in the morning. Michael was out of town, but would return the following day.

I checked in. All the hotel signs were in Pidgin. On the inside of my door, a sign hung hooked around the door handle. "Yu No Ken Kam Insait" (from the English, "You no can come inside," or "Do not disturb"), it read on one side. "Yu Ken Stretim Rum Naw ("You can straighten room now"), the other side said. I left and walked outside. I had forgotten the degree to which mountains ringed the town on all sides, grazing the clouds. Yet the town itself seemed poorer. The commercial strip hadn't grown at all, and in fact had retreated. Only one new building was under construction (by the Asia-Pacific Company), and a few older stores had closed. The snack bar where the Australian once snapped a picture of me was now boarded up—gone. Barbed wires encircled the police station. The lot where I learned to drive a stick shift was overgrown with chest-high grass and plants. Now, in the dry season, heat and dust hung over everything. Along the side of the road, silvery grey chalk coated red and green leaved bushes. Packs of barefoot men squatted in the back of rickety pickup trucks—PMVs. The town was more crowded and noisy than before.

The market hadn't changed much. Women still sat on the ground behind piles of coconuts and dirt-caked *kaukau*. In the hot sun, people wore headgear—hoods of winter coats minus the coats, clothes over their heads to keep the sun out of their eyes. One woman wore a yellow plastic bucket as a hat. To escape the heat, people sat in the shade of fences and crowded at the base of the two tall pine trees beside the airstrip.

I was the only Caucasian walking around, and the only person wearing shoes. Everyone else walked barefoot. People caught my eye and nodded. Children stared—to see a White person was still something of a novelty. They caught my eye and lit up. When I smiled back, they became embarrassed. They still had a sense of wonder here.

Back at the hotel, I had to wash a coat of brown dust off my face and hands. More New Guineans were eating at the hotel restaurant than would have in the past. Some of the population obviously had money. Still, the waiters were barefoot and clad in wraparound skirts. Their pens stuck up vertically from the back of their hair as *bilas* (decoration). I ordered the "Reef and Beef Buffet." When I walked back from dinner, the night was lushly fragrant, scented by flowers in the dark. Yet in most of the rooms I passed, expats sat watching TV.

I was back in my room when there was a knock at the door.

"Yu likim mi turnim down the bed?" ("You want me to turn down the bed?"), a short man asked—from housekeeping.

"Sure."

Exhausted from traveling, I soon lay down to go to sleep. Suddenly, a mosquito bit me. I got up, turned on the lights, searched, found, and killed it. I went back to sleep, but got bitten again. I turned on the lights, stood on the bed, scanned around, spotted the bug, and slapped it against the wall with my notebook. I lay in bed again, only to be attacked once more. This happened seven more times during the night, hampering sleep. When the sun came up, a mosquito sat on the inside of the mosquito mesh, trying desperately to get out. Dead mosquitoes splotched the back of my notebook.

As I was taking a shower, birds'outside whistled, twittered, and warbled—a chorus of flutes questioning and answering each other. Some each sang several notes—four long, mournful notes, then in another voice, four short, rising ones, as if different birds. The music awakened and refreshed me.

I went downstairs for breakfast—a buffet of cold cereals with local fruit. Fresh pineapple, mango, and papaya tasted a little sour with milk, but added texture to the cereal. The only other guest was an Australian businessman "in transport." We sat down together. "The country's going backward," he told me. "On the Highlands Highway, the paved sections got worse than the dirt ones because of huge potholes, so they're breaking up the asphalt now to turn it back into dirt. In the past fifteen to twenty-five years, the population has doubled. The number of expats has halved. The politicians are all totally corrupt. Everyone in the government from the top down skims off the top. If they don't, everyone thinks there's something wrong with them. The Prime Minister, Julius Chan, is now worth 100 million kina, and started with next to nothing. He owns Island Air—the second-largest airline in the country—and is a major share holder in some of the mines. Members of Parliament each get a 300,000 kina a year slush fund to use for schools, clinics, roads, and other services in their local districts. Instead, they use the money for election bribes and to fill their pockets. Put up 1000 kina to run, and if you win, be a millionaire in five years—not a bad investment."

"What's it like to live here these days?" I asked.

"Well, it's exciting, never routine. Besides, you don't get this in Adelaide," he said, pointing to the terrace of potted purple, pink, yellow, and orange flowers. "We do things to keep each other sane."

"Like what?"

"Well, we play a lot of party games."

"Party games?"

"Yes. In one, we go to each other's houses, and walk from one end to the other without touching the floor—walking on the furniture, hanging from the drapes. In another game, people shit in the holes in the one golf course in town. In the morning, when the golfers put their hands in the holes to get the balls . . ."

I was glad to head off to the Institute to visit Michael. His beard had grown all white now and tufts protruded on either side of his head like mutton chops on an elderly nineteenth century English philosopher, scientist, or writer—a Michael Faraday or John Ruskin. We talked about what happened to Carleton, whom I had seen a few months before. Then we discussed Waisa.

"Two British researchers are there now," he told me. "John Collinge, an English professor of neurology, is here just for two weeks. Jerome, a nurse who works for him, is drawing bloods from the Fore and has been there for six months. Before Jerome arrived, the house had been unoccupied for eight years."

"How are people reacting to the fact that it's the end of the epidemic?" I asked.

"Don't talk about it. It might come back." I couldn't tell if he was describing his own attitude or that of the Fore. Perhaps both. "The incubation period could be as long as sixty years, though I won't be around to see it. I'm going back to Australia next year. Then, the epidemiological study will stop. There will be no one here to carry it on." It sounded sad.

Michael arranged for a driver from the Institute, Steven, to take me out to Waisa the following day. As I left Michael, Koiya, who watched Carleton's house in Agakamatasa, walked by. Recently, he had been accused of being a sorcerer, of causing kuru and *tukabu*, poisoning people. As a result, he had been attacked—cut in the neck with a knife and almost beheaded, and stabbed multiple times in the chest. He had to flee Agakamatasa to live in another tribe in another area, where he had been ever since. The neck wound had healed as a thick, dark, raised five inch-long gash. I felt very sorry for him. Superstition still thrived.

At the hotel, I packed a knapsack to take. I would leave the rest of my luggage in Goroka. I realized I had brought what might be the first Barney's suit to New Guinea—having taken it to the Australian conference—the first Zabar's bag, the first Clinique products.

I also went food shopping. The grocery store had shrunk its selections, no longer stocking Fortnum and Mason tea, every kind of Twinings tea, and gourmet cheeses. I was afraid to buy "Ox and Palm" canned corned beef, even if "packed according to PNG Government Inspection." But I purchased pasta, bread, cans of tuna fish, salmon, oysters, and mussels, and a tin of margarine. I wanted to buy a notebook, but had difficulty finding one for sale in town. Gone was the store where I once took my watch to be repaired. In its place a store sold a hodgepodge of cheap clothing and household items. Flat on a shelf lay six dusty used books—two on New Guinea, a copy of *Tok Pisin*, and three children's books. The six books occupied five feet of shelf space. The store sold no 8½ × 11" notebooks, only a small stenographer's pad.

I now found myself apprehensive about going for five days to Waisa—somewhere that lacked electricity or phone. I wondered how many such places still existed. Before I left, I decided to call my office in New York to tie up a few loose ends at work. I tried calling. But I couldn't get an outside line from my hotel room, and called the switchboard. "The phones are down," the desk clerk told me. "I'll try getting an outside operator. What number do you want to call?"

Half an hour later, he called back. "The international operator said that was a nonworking number."

"It works. Please try again."

"No, he said it doesn't work."

"I know it does!"

Finally, I got through, though there was a several-second delay between everything said, such as used to be the case with all calls between North America and Europe, but which advanced telecommunications has essentially gotten rid of.

In the morning we drove off. Steven stopped by the shops. Forty Fore jumped on the back of the truck—adults and children, carrying *bilums* of *kaukau*, four-foot-high sugar cane stalks, and live chickens.

"You're the one who drew the picture that's hanging in Waisa!" one man said.

"You know that picture?"

"Yes." He couldn't believe it—that I looked like the drawing.

We stopped again just at the edge of town at a market now devoted solely to betel nut. Two lines of women sat behind piles of green fruit that resembled tiny limes. "Em i got bigpela tru na emi i got liklik," Steven explained ("They have really big ones and little ones"). Each woman sold a different size. He also bought strips of a fruit called *dagma* that turned his mouth red. Women, I later heard, traded sex for betel nut on the coast, then sold the betel here in the Highlands. Unfortunately, the AIDS epidemic was rapidly increasing here. These women were at risk.

Immediately out of Goroka, beyond a few streets, we hit bush—little development, groves of bamboo, dry dust. Suddenly, only minutes out of town, the asphalt ended and turned to dirt.

We started to climb the hills, and the valley soon swept below us. The ranges seemed taller, vaster than I remembered. Before, I had focused on what was immediately in front of me—people and interactions—rather than on the peaks against the sky. The mountains' arms swept down one after the next, their spines running together and disappearing into the sky. Behind soaring ridges others loomed. Even in the gaps, further crests rose. Summits emerged through the clouds, with other peaks hovering above, topped by white mists, and then even higher ranges. Many mountains were taller than they were wide. Some were almost vertical walls, and seemed too close to be rising so high. Usually such heights lie at more of a distance. Only after I was last here had I seen other great ranges—the Rockies and the Himalayas. Yet those around me now were swifter in their ascent, more unrelenting. Since I was last here, I had developed and looked at photographs I had taken of this area. But they lacked the sense of depth, the shortness of distance, the spaciousness of the terrain itself.

In the truck, the smells of the Highlands again came back to me—wood smoke, dust, and body odor. These were a people without antiperspirant, showers, or shampoo. Little had changed. Banana trees lined the road like an avenue of landscaped shrubs in other parts of the world. A wild boar calmly crossed the road. Last time I was here I had stopped seeing many things afresh or thinking about them and had eventually taken them for granted. Yet to return to a place where we once spent time jars our memories. Images came back to me, reconnecting me with the past. I was traveling ever backward over millennia in human cultural development and years in my own life.

In the five-hour drive we passed only two other vehicles—both flatbed trucks jammed with people coming in for *singsings*, celebrating election victories. The women stood in front with café au lait clay powdering their bodies, their unclothed breasts hanging, swinging, their arms raised high, throwing fists into the air. The men had garlanded branches of leaves into their hair. The day before, a village had been burned down because its candidate had won a Parliament seat over the candidate from a neighboring village. Two nearby villages with opposing candidates each said they would torch all the bridges on the road if their candidate lost—in either case the bridges would be destroyed. Elsewhere, a woman carrying her three-week-old baby in a *bilum* was raped by forty men, her husband axed, because someone from her hamlet had won. Everyone felt the election was important. Yet there were no issues, just greed and eagerness to win because of the benefits to a winner and his *wontoks*. People treated the election as tribal war.

The road had deteriorated. The Australians, when they built it, had maintained it, keeping up a fleet of bulldozers to plough and smooth it out periodically. The bulldozers had remained in use when I was last here. But now, we passed one or two rusting by the side of the road, broken, and left to rot. One- to two-foot ditches gouged the road vertically and horizontally. Steven drove less than ten miles per hour. Still, we bounced up and down, side to side, moving not only forward but up and down. My head felt like that on a toy china doll, attached by a spring, flopping back and forth, side to side without stop. "Road" was a euphemism. It was more a long strip of dirt, the width of a car, with no trees growing on it.

Slowly, we climbed steep Kuru mountain. From the top, even more came back to me as I now peered out over the hidden valleys of the Eastern Highlands. I realized how much kuru had spread here because of these steep ranges and the isolation they created.

Along the road, old men still carried big loads—long fronds taller than they were, for *mumus*. Women hauled huge *bilums*, heaped sacks of rice and yams on their heads, and babies and other loads in their arms. Even young girls carried small babies, supported on hips.

A few houses were different than before, now built on stilts with narrow ladders of rough-hewn wood leading up to six-inch-wide terraces, imitating houses in town and on the coast. Fashions thus change—even here. The quest for new fashions thus seemed a basic element in human nature. Evolutionarily, this search for the new no doubt confers a selective advantage, helping to locate new and potentially beneficial practices and behaviors.

Yet little else here had altered.

Suddenly, rattling over the planks of a bridge, I felt a jolt. A few yards later, the truck ground to a halt. Steven got out. "Spring em i bruk. Em i bagarap" (the spring or shock absorber had broken, "buggered up"). We all climbed out. Steven found the rubber and the metal washer and replaced them. We piled back in and continued on.

A few minutes later, we felt another pop, and got out again. This time, Steven found the rubber washer, but not the metal one. We all scoured the grass. I looked, but didn't find it either. If they couldn't locate it, I thought I certainly wouldn't be able to. After forty minutes, Steven found a bottle cap, pulled out a bush knife, dug a hole in the cap, and used it as a replacement for the metal washer. Twenty minutes later the truck broke down a third time. We all stood about for another hour as Steven fixed it.

We passed coffee plants—more and taller than before—growing as if weeds in cool groves of eucalyptus-like trees.

Then we passed the turn off to Purosa, and neared Waisa. This was a road that I had traveled many times on foot. It felt odd coming back. I had never imagined I would one day return.

There, before the expanse of mountains sat Soba's small house and ours. All seemed so quiet—once the scene of so many confused thoughts and emotions. Smoke rose from the fire in a nearby hut.

People came up to me and shyly shook my hand, saying "apinun," from the English, "afternoon, or "good afternoon." More people now spoke Pidgin. Still, none spoke English.

A young man walked up to me, wearing sunglasses and a brand-new rugby shirt. He seemed happy to see me. It was, I suddenly realized, Jason.

"I named my son after you," he told me.

"You what?"

"Klitzman, come here," he said to a short, 4-year-old Fore boy wearing a long, dirty brown shirt and nothing else. He was barefoot. "This," he said to me, "is Klitzman. Klitzy, come here." Klitzman ran behind his mother's legs.

Jason shook my hand further. "Aiiiii! Aiiiiii!" he sort of laughed, in utter disbelief that I was here. "Remember me going with you to Tagowa and Mentilasa?"

"I do," I said.

"Are you married now?" another man asked me.

"No."

"You were really a *pikinini* when you were here, weren't you?" he said—a child.

We talked a little longer, but I was standing with my bag and a box of food I had brought. "Let me go into the house for a minute," I said, "to drop off my things. Then I'll come back."

I walked up to the porch.

Gone were the passion fruit and flowers we had planted in front. The lemon tree had been chopped down, probably for wood. The outhouse had been moved a few feet over.

I entered the main house. It was much smaller than I remembered. Memories suddenly came back to me of just relaxing during the days when I didn't have much to do and could stay here in the cool mornings by the window. In all the difficulties of being here, I had forgotten the few pleasurable moments—being able to sit for hours, drinking tea, eating tomatoes or toast with jam for snack, and occasionally listening to Roger's short-wave radio. My picture and Jason's still hung on the wall, taped up amidst New Guinea artifacts that had been put there over the years. It was, I realized, a portrait of me as a young man—earnest and innocent. I was surprised how young I looked. I had the whole world ahead of me then—though at the time I didn't fully feel it. Rather, I thought I needed to be careful, make the right decisions, be mature. I realized how much I had changed since last here. I was now a physician, a psychiatrist, and an assistant professor at a medical center, and had written a few books. I felt more confident about my reactions, intuitions, and ability to act on them—all of which I had just begun to do here.

My room now looked smaller to me and resembled a dorm—narrow with a bunk bed and a tiny desk at one end, underneath a window. I didn't see it as such at the time—I had just graduated college and was used to such accommodations. The desk in the room barely fit a typewriter, but had done that. I'd never thought I'd see and touch the table again. The surface was uneven, I now noticed, made of two parallel boards, one a little higher than the other, both covered with slightly bumpy yellow and green linoleum—a design of floral Delft-like tiles.

Everything was exactly the same—the pots and pans, the food cupboard with metal mesh, the flat metal sink counter in the back where I washed up. The shower room was too small to move in, only for standing still. I had forgotten the rich, sweet, woody smell of the house. I felt I was in a dream—a place familiar not from the present, but the distant past. At the time, this seemed the remotest part of the inhabited earth, certainly among the most isolated—it still was. Yet Roger, Maryanne, and I had long since gone, leaving nothing behind except their washhouses and my drawing. It was an odd and disturbing premonition of growing older and eventually departing from the earth entirely—a glimpse of the world in which one no longer exists. Mostly, though, I was just amazed to be back—this place, so far away physically from my current life, was still here, as if a time capsule, almost as if I had left it yesterday. I felt I had changed more than it had. As I looked outside, memories of other things returned, too—the tall *kunai* grass and sugar cane, the live fences—things I had seen only here.

Yet the village somehow seemed even poorer. Coffee doesn't grow near Waisa as elsewhere along the road because of soil or climate. The clothing was raggier, dirtier. Pigs had chewed and dug up the compound's entire yard, making one endless stretch of loose soil. I had had to step in piles of earth and holes. Dirt coated everything.

Jerome and John Collinge were out hiking to Purosa, but soon returned. John was tall and spoke with a distinctive English accent. Jerome had on a well-worn cotton shirt and hiking shorts. He had bouncy, bushy brown hair and wore wire-rim glasses. We sat down together. I took my old seat on the right by the window, and asked about their research.

"The United Kingdom is about to experience an epidemic very similar to kuru," John told me. "Twenty people have already died of CJD in Britain. Many more will. If more people in Britain knew someone who died of CJD, they'd be much more worried about it. Kuru is also important for understanding cancers. Oncologists are suddenly interested in infectious proteins. Oncogenes like P50 are isomers of cancer suppressor genes—that is, prions—and when present, cause cancer. Undoubtedly many other human infectious protein diseases exist. Kuru is just the first one found."

"Yet there is much still to learn about it here. We need to find kuru's mean incubation period, which I think we can by finding out when the sex ratio changed to one to one—that is, when only the children at feasts were affected. That would help tell us when to expect a major epidemic in Britain, or whether the short incubation periods we've seen as of now are the mean, and that we're thus now in the middle of the human epidemic, and it's small."

"But," I said, "I found that most people were at multiple feasts. It will be difficult to figure out the point of exposure, as there were usually several."

"Your research sounds extremely relevant. I would be very interested in seeing it."

"Why there's a copy of your published paper here," Jerome inter-rupted. Sure enough, the paper I had written in Madang and published was in a box of information about kuru kept at the house. It was, I sup-pose, what made it all worthwhile in the end, bringing my work full circle. I also realized I could go back and look at the original genealo-gies I had collected to see what they suggested about mean incubation periods

John took the paper. "We also still don't know," he continued, "the structure of the host protein in the case of kuru and whether different strains exist. The CJD we're seeing transmitted from cows appears to be a new variant molecularly and clinically, related less to scrapie or other CJD than to BSE. In other words, I don't think Mad Cow disease came from sheep, but from cow offal—cattle with a primary infection. It would be important to know how kuru compares."

"I wonder," I said, "what if kuru came not from a case of CJD as is commonly assumed, but from another mammal—a possum, pig, or rat—that had an infectious protein disease?"

"That is a very interesting question," he said. "Especially as we're now finding that people's susceptibility to BSE varies, it's important to know how vulnerable the Fore were. We've done DNA tests and found that there are different strains of the host protein of which the agent is an isomer. In other words, some people in Britain are more likely than others to get CJD from cows. We now have newly developed DNA tests to determine homologous vs. heterogeneous DNA typing of the normal prion gene, predicting susceptibility. People might differ in susceptibility to kuru as well. If the Fore are more genetically suscep-tible, less people might die in Britain than previously thought, based on the model of kuru. As a result of their genetics, patients with long versus short incubation periods may also display different symptoms. We need to see if the clinical picture is now changing. All this would be important to know. Jerome is drawing blood from everyone. But if only we could see and get bloods from *patients*, we could answer a lot of major questions: how many people will die of BSE in Britain. But we are too late. The people here now were not exposed, and unfortunately there aren't any cases."

"But there were cases when I was last here. Also, some people were exposed and didn't become sick," I said. "I collected their bloods when I was here before."

"Really? Why, those bloods are invaluable! We could run our new tests on them. If these people do have genetic patterns of lower susceptibility, it could also help in developing treatment. Where are the bloods now?"

"I don't know. But we can check with Michael."

I wished I had collected blood samples from the kuru patients I had seen and those with differences in incubation periods. But I hadn't. Still, I had some bloods. I also realized that the wide variation in incubation periods I had found among the three age mates in one cluster I identified might indeed be due to inherent genetic differences—not dose or route of transmission—further implying that some people in the United Kingdom and elsewhere may be more at risk of CJD than others.

That evening, as light disappeared behind the mountains and crickets buzzed and sang in the darkening night, Jerome filled me in on things that had happened here since I had left. The Lewises had long since returned to the States, but had first built a larger missionary school and clinic up the road at Ivangoi. The library had moved to Ivangoi, too. The people of Waisa were upset, but no one had used it much.

Jason had returned not long ago to the village after Michael had sent him to college on the coast. Jason hadn't tried very hard and had dropped out. Recently, Sayuma had been sick, and instructed Jason to get money from a candidate who had promised Sayuma 300 kina as an election bribe. Jason got the money, but kept it, and bought clothes for himself in Goroka.

John took out a cellular phone he brought Jerome to use here. Satellite communication had indeed now arrived before electricity and telephone wires. "Gee, now we can order for takeout food," I joked. John and Jerome both laughed. It felt good to make light of all the difficulties here.

In the morning, we went to see the one rumored current patient. Jerome had gone three times before and been refused each time and told to come back. One time the man and his wife hurried off to hide in the bush when Jerome was seen approaching. "This will be my last attempt," Jerome said.

Jerome's three new kuru guides—Jim, Sana's adopted son, who remembered me staying in his house, Minari, and Camber—drove with us to Purosa.

"What do you think of kuru ending?" I asked Jerome's guides.

"It hasn't ended."

"It has gone down though."

"That's because the sorcerers have mostly died. But there are still other forms of poison and sorcery—*tukabu*—that go on." A baby had recently died—of poison, it was decided. Three elderly women, said to be responsible, were hanged and burned. The smoke could be seen for miles around. "People get attacked by sorcerers on the road at night now. The sorcerer then uses poison to make the people forget they were attacked. Later, the sorcery takes effect in the body."

Jerome had been told to go to a hilltop near a new missionary church. We parked at the base and walked up. There, on top, three men sat on a bench near a few dark, windowless bamboo huts. We shook hands. The oldest man, at the end, seemed a bit shaky. His hand trembled slightly as he shook mine. Discreetly, I watched him. His head and legs moved about a bit, almost restless. I could tell something was wrong. He had kuru.

Jerome's guide, Jim, spoke to the three men in *tok ples*. I walked up to John. "The man on the left has a mild tremor," I whispered.

John looked. "You think?"

"I do. I think he's the patient."

"It does look like he's restless and trying to control it." But John wasn't yet sure.

I had seen a lot of kuru—more than John. He had examined cases of CJD acquired from Mad Cows in Britain. But I had seen about as many cases of kuru as there were cases of cow-acquired CJD identified at that point in the world.

"Okay," Jim said after speaking to the two younger men present, "you can see the patient. It is this man here." Sure enough, it was the older man, named Wanupo.

We asked him to stand up. Then we asked him to walk heel to toe. He tried but stumbled. His right foot stamped to one side. He stopped. He couldn't do it. A crowd had now gathered. We asked him to try touching his finger to his nose, and other coordination tests. But he couldn't. He thought he had malaria, and complained of the top of his head being hot. "We should bring the case for Michael to see, to make sure," John suggested.

"No," I said. "I know. It's kuru."

I felt very sad for Wanupo—sadder than for any other patients I had seen before here. In part, since my last trip here I had seen many more patients. In the interim I had gone to medical school and become a physician. Through internship and residency, I had treated patients with various ailments over weeks, months, years. I had gotten a fuller sense of what it's like to be a patient—the on-going suffering. I didn't just come and go as I had with patients here before. I was no longer just a researcher, but a clinician and healer. Previously, I had felt badly for the patients here, but also helpless. Now I had seen more clearly what could and could not be done—how, though we couldn't always cure, we could provide diagnoses, prognoses, explanations, sympathy, and understanding—often much appreciated.

I asked him how old he was. He didn't know. Nor did he know when he was born. "Was it before the road to Purosa was built and John James came?"

"Yes."

"When the road was built, were you the age of this boy?"—an infant—I asked.

"No."

"How about this one?" A 7- or 8-year-old.

"Yes."

I knew the road was built in 1957. Thus he was born in 1949 or 1950, making him 46 or 47 years old, and giving him a probable incubation period of at least forty-one years, as he would have been an infant at the feast and the feasts ended shortly before the road was built. His mother, too, had fallen victim to kuru. He would now die from a meal consumed forty-one years ago. As an infant, he had had no choice about participation and besides, no one knew of the danger. He had lived all this time, wholly unsuspecting that a time bomb was ticking inside him. I suddenly realized the implications. It was sad, not just for him, but for Britain and the world. People would still die in Britain of Mad Cow disease more than over forty years after they ate their last infected hamburger. In other words, through the year 2036, if not longer! This man had the longest documented incubation period in a human being, or any other species. Here in this village, with its dirt, bare huts, and unwashed feet, a scientific record was being set. Similarly, in Britain now, thousands probably harbor the agent unknowingly, unaware whether or when it will kill them.

The disease had not disappeared. The epidemic was not over. Some day there will be a last man or woman to die of kuru—the one with the longest incubation period (though no one will know it at the time). It was too early to assess if that had yet happened. Moreover, some may die of other causes, still incubating the disease, perhaps even after sixty years. Why had he survived over forty years, while others had survived only a few? Did his genetic susceptibility differ from those who had died earlier? We now hoped to be able to find the answers.

It was hard to get additional information on his genealogy and on what feasts he may have attended. Wanupo himself was one of the oldest men around. (Life expectancy here wasn't high. Many suffered from infectious diseases and lung disease acquired from the constant inhalation of smoke in closed huts.) I had been here during a unique window in time. There had been individuals who had been to only one or two feasts, and older villagers who remembered the time of cannibalism.

"We need to try to get an autopsy on him and get his brain," John told Jerome. "Do you think you can perform it here?"

"Yes," Jerome said.

"Good." The conversation made me uneasy—this eagerness to obtain a cannibal's brain.

"Why do you need an autopsy?" I asked.

"To investigate where in the brain the agent goes, what the molecular structure is, how close it is to CJD, and many other scientific questions we couldn't even imagine a few years ago." The brain would indeed be important.

We climbed onto the truck and drove back in silence. At the house, Soba came in. He still wore a woolen ski hat as he had 16 years earlier. Even here, with limited resources, people maintained personal styles and trademarks. At first he ignored me and only later said he remembered me. I asked him how he was. "I am old," he said. "I don't do anything." He smiled. "I let everyone do things for me." He said he had a cough and asked us to examine him. He seemed okay, and then told us he wants a *haus sik* built here.

"Do you remember Roger and Maryanne?" I said.

"Yes." But he would say nothing else about them.

"There must have been a lot of changes since then."

"Old people have died. New children have come up. Services and the road have gone down."

"Have there been many researchers since I was here?"

He shrugged. "One doctor drank too much. I told Michael to get rid of him and he did."

"Really?"

"Yes."

"We here in Waisa made Michael's career, you know. Because of us, he's doing well for himself in Goroka—*bossman* of the whole Institute." Soba was the same—shrewd. He wasn't going to talk much, but wanted to sit and didn't want to leave. To sit in the house was important to him, and conferred status. Our house still had the only table and chairs in the village. John wanted to talk to me about research, and started to. Finally Soba said, "Okay, yupela tok *tok ples* bilong you" ("Okay, you can talk your *tok ples* or local language"). English was seen not as the main root of Pidgin, but merely as another *tok ples*—one of many.

Outside, George—Soba's adopted son—and his *wontoks* prepared a *mumu* for us. They had started in the morning. A dozen women sat all day peeling a huge three-foot-high pile of taro, a large white tuber. "They're making a mumu WITH LAMB FLAPS!" Camber and Minari told us, excited and amazed. Lamb shanks—greasy fat attached to ribs, fed only to animals in the West as rejected meat—were considered an expensive delicacy here. At the *mumu*, George proudly set before John, Jerome, and me a big bowl of the steamed greasy meat and vegetables— mostly bland, pulplike taro. The meat was poorly cooked, oiling up my lips, nostrils, and fingers. Everyone looked at us, motioning for us to try it. How was I going to get through this without offending them? I took a bite of the taro. It was like paste, but I forced myself to chew a mouthful and smile. "Don't mind my fingers," John said. Fingers were all we had. Luckily, the corn and *kaukau* were good—familiar to me. George also served mushed bananas with lamb flaps cooked in banana leaves, which also greased up my mouth and made my stomach queasy. A baby pig, its fur cream-colored with two or three big brown spots covering half its back, wandered up and poked its head into our bowl. John waved his arm in the air to chase the pig away. The animal scurried off, but slipped up to another pile of food not far away. Someone else soon pushed him away again.

"This is all kind of gross," I said to John under my breath.

"You know," he replied, "pigs are as smart as dogs."

"Really? Then how come they're pigs?"

The little animal sidled up to yet another pile of food. Suddenly, a girl grabbed him, and carried him off as he cried and squealed loudly. All of us at the *mumu* laughed.

That night, John, Jerome, and I were still hungry and cooked dinner for ourselves. In the dark, I scrambled eggs for us on the wood-burning stove. I couldn't see when they were done, and had to lift the pan and bring it under the Coleman light on the table. I felt I was camping out.

Jerome's guides, Minari, Jim, and Camber, came over. "We want the Institute to build another house," they said, "on neutral ground, as a 'clinic' or *haus sik*. You shouldn't be here," they said. "This property belongs to just one line."

"But the Institute already has a house here," Jerome said.

"Well, okay. You can live here, but not work here, too. Besides, it's not good for blood and food to be at the same table."

Jerome said he'd talk to Michael.

"And," Camber added, "we want a pay raise." Indeed, little had changed here.

At night, dirt and dust coated my clothes, my face, my hands. I decided to take a shower. Jerome never used the shower room. "It's not worth the effort," he said. He washed up in the sink instead, but I longed for a shower. The bucket was on the ground, filled with cobwebs. It hadn't been used in years.

The next day we got ready to leave for Goroka. After a few days here, the house seemed bigger again, comfortable, efficient with its specialized rooms, nooks, and crannies. I packed my bag and walked outside. I was surprised that a crowd had gathered in front of the house to see us off. I approached Jason. "Be careful," I told him. "You're smart. Don't give up on doing something with your life. But you don't want a bad reputation. Don't be a rascal." He knew what I was talking about, and got embarrassed. I think he was surprised that I was aware of the problems into which he had gotten himself.

At the truck, dozens more showed up to say good-bye—the entire village. Soba's adopted son, George, walked up to me, "Waisa em i as ples bilong yu" ("Waisa is your *as ples*"—from the English "ass place," meaning the place where you sit—your place of origin, where you're

from). "England em namba tu. Tasol Waisa em i nambawan as ples bilong yu" ("England"—where they now assumed I came from because
John and Jerome did—"is your second place of origin. But Waisa is your
first"). They now thought New York must be a part of England.

George's wife gave me a beautiful *bilum* she had woven. *Bilums*
were now made of brightly colored commercial yarns. The art of making these bags from bush rope and native dyes will undoubtedly soon
be lost . Camber gave me arrows he had made. Jason gave me a long,
black bow carved from a special palm tree. "Put them up on the wall of
your house," Camber said. "Bilas bilong haus bilong you." They saw
houses, not just people, as having *bilas*—decoration.

We drove off slowly, back down the bumpy road.

Back at the Institute, Deborah called me over and introduced me
to a tall, strapping young man with a beard. "Remember Binabi?" she
asked. I couldn't believe it. He, who had lived with the Alpers as a boy,
was now married and had two children of his own.

That night, Michael and Deborah met John, Jerome, and me for
dinner at the Bird. I wore a plaid shirt I had bought at Banana Repub-
lic, old khakis, and a belt—my least nice clothes in New York. Yet I felt
overdressed. People had few clothes here and wore them a lot.

I asked Michael about the bloods I had previously drawn. He would
track them down. They would now be used, and might offer clues.

At 9:50 p.m., Michael and Deborah abruptly got up to go. Several
other people quickly scurried out of the restaurant as well. Some New
Guineans remained seated around us, drinking. "They'd better be staying here at the hotel," Deborah said.

"Why?" I asked.

"The 10 p.m. curfew"—a reminder that trouble still lurked.

The next day, men rode into town, standing in the back of packed
trucks, wearing leaves in their hair, waving branches, and shouting
two-note chants, "Ah! Oh! Ah! Oh! Ah! Oh!"—victory parades. It was
announced that the Prime Minister, Julius Chan, had been ousted.
Everyone was excited about the results: new people voted in, the old
ones out. But the system remained unchanged.

The next night, my last before I was supposed to leave, I had wine
at Michael's. I saw again the kitchen where I had cooked, where it was
possible to cook with spices, refrigeration, and a wider range of ingredients than at Waisa; the glass-fronted cupboards of wine glasses that

had all rattled during the earthquake like a seismograph; a complete set of Shakespeare plays; the complete *Rise and Fall of the Roman Empire*. I had forgotten the low-lying modern chairs flanking the corner table where under a lamp's pool of light a phone sat—on which I had called home when I first arrived.

The next day, one of the Institute researchers, Patrick Miese, drove me to the airport to catch my flight. I was headed to Moresby where I would connect to a flight to Australia.

"I'm sorry," I was told at check-in. "But we don't have any planes today."

"What do you mean?"

"The flight from Port Moresby never arrived so we don't have a plane to go back."

"What am I supposed to do?"

"We should have a plane here tomorrow."

"You *should?*"

"Yes."

"At what time?

"We don't know yet. Come by in the morning."

I couldn't believe this. It wasn't clear I'd be able to make my connection tomorrow either. Pressure and frustration boiled up within me. Patrick didn't seem surprised. "You thought there'd be a plane?" He laughed. "These things happen here all the time. Come back for a drink." I went back to his place. His New Guinean cook made us drinks. The cook then sat watching TV films with English subtitles. He was illiterate and spoke only *tok ples* and Pidgin, but sat mesmerized in front of the TV.

First thing the next morning, I headed to the airport. A few hours later, a flight arrived from Port Moresby and I boarded the plane to return. Most of the passengers were nationals and barefoot. We flew into the clouds and an hour later landed in Moresby. Baggage Claim consisted of one man throwing the luggage from two trucks onto a table. Bushels of bananas had been checked on and tagged as luggage. Printed destination cards were tied to the stems with tiny white strings. I retrieved my bags and caught a flight to Cairns, Australia. The passengers now were exclusively White. As we took off, I glanced at New Guinea, now disappearing on the horizon. Travel in this country was even harder this time than before. Phones, planes, and roads had all deteriorated. Yet

I was very glad I had returned. New Guinea had remained unique, a natural laboratory for understanding people, culture, and the effects of time. I thought it would be important to return again some day, but I didn't know whether or when I ever would, and if so, how much it or I would be altered. Still, I had learned a lot both this trip and last.

I realized how significant my experiences in this country were, how conventional and predictable those since—in medical school, internship, residency, and fellowships. Travels during vacations hadn't come close to challenging me in as unexpected ways. My time here, I realized, marked the beginning of my adult life. It was my first job after college, and the most adventurous thing I'd done before or since.

An hour and a half after leaving Moresby, I landed back in civilization. I was surprised how relieved I was to be reconnected to the rest of the world by e-mail, fax, and modern telecommunications. I saw how much I was part of my own culture—even with all its excesses and flaws. That evening, I stayed in a resort on the Australian coast and had dinner by a pool, surrounded by trees. New Guinea was a short flight, but twenty thousand years away.

Camus ends his novel *The Plague* with a doctor wondering when, if ever, the pestilence will return to a town. It is hard to know when the end of an epidemic has occurred. Any new case marks a continuation. Still, two generations of Fore have arisen since the numbers of kuru cases began to decline. Others come along who never knew the epidemic. It recedes ever further into the past. Yet fears linger much longer.

In coming years, other plagues, caused by new, poorly understood infectious agents will no doubt burst forth. Formerly isolated regions of the globe interact more and more with each other. With increased speed and frequency of air travel, we can leave a tropical jungle in the morning and be in London or New York that night or vice versa—along with the parasites we harbor inside us. In this way, over recent decades outbreaks have emerged of Ebola, Lhasa fever, and Legionnaire's disease. New niches for parasites continually form. Increasingly, we are all both vectors and potential victims of previously unencountered infections. Such new plagues will seriously challenge our knowledge of diseases, epidemics, and human beings.

When I started medical school, the field of infectious disease had lost prestige and a sense of excitement. Physicians had assumed that contagious diseases had been conquered. Yet "old" ailments such as tuberculosis have reemerged and, along with these assorted "new" outbreaks, reveal our earlier hubris.

The story of Mad Cow and other infectious protein diseases is far from over. A previously unknown one seems to be discovered every few years. Countless more human cases may yet appear, perhaps over many decades, both in Britain and elsewhere. Epidemics that arise of these and other agents will no doubt prompt fear, resistance, and denial—abetting the further spread of disease—and desperate searches for answers and solutions, as I saw in New Guinea.

Index

Acknowledgments

I am indebted to many people and institutions for their help wth this book—most importantly, Dr. D. Carleton Gajdusek, who sent me to Papua New Guinea, and the National Institutes of Neurological and Communicative Disorders and Stroke, under whose auspices I worked there. Michael Alpers oversaw my work at the Papua New Guinea Institute of Medical Research. He and his wife, Wendy, provided hospitality in their home, without which I would not have survived, during my stays in Goroka. I also wish to thank Graham and Rosalyn Henderson, Deborah Lehmann, Auyana, Anua, and Igana; and the patients and their families whom I interviewed.

For help with this manuscript, I want to thank Richard A. Friedman, Renée C. Fox, Shirley Lindenbaum and particularly Royce Flippin. I am very grateful to my agent, Kris Dahl, and her assistant, Kimberly Kennerly, and to my editor, Erika Goldman, for her help and enthusiasm. For assistance typing the manuscript, I want to thank especially Royce Lin, and also Nancy Lange, Roberta Leftenant, Rohit Bansal, and Blake Brinson. For the opportunity to study cultural issues in epidemics, including AIDS, I am indebted to the Aaron Diamond Foundation, the Robert Wood Johnson Foundation, the Picker/Commonwealth Scholars Progam, and the National Institutes of Mental Health (through career development grant K08 MH1420-01, Center grant P50 MH43520, and fellowship training grant T32 MH19139). I am also grateful to the Corporation of Yaddo, where I wrote part of this manuscript.

About the Author

Dr. Robert Klitzman, currently an Assistant Professor of Clinical Psychiatry at the College of Physicians and Surgeons of Columbia University, is the author of *A Year-Long Night: Tales of A Medical Internship* (Viking, 1989) and *In a House of Dreams and Glass: Becoming a Psychiatrist* (Simon and Schuster, 1995), and *Being Positive: The Lives of Men and Women with HIV* (Ivan R. Dee, 1997). He graduated from Princeton University and Yale Medical School, and has been a Robert Wood Johnson Foundation Clinical Scholar at the University of Pennsylvania. His work has appeared in scientific journals and textbooks, as well as the *New York Times* and other publications. He has received numerous honors and awards, including a Burroughs Wellcome Fellowship (for Future Leaders in Psychiatry) from the American Psychiatric Association, an Aaron Diamond Foundation Fellowship, and a Picker/Commonwealth Scholar Award. He has also been a Fellow at Yaddo and MacDowell.